GE 140 .V58 2013

Vital

About Island Press

Since 1984, the nonprofit Island Press has been stimulating, shaping, and communicating the ideas that are essential for solving environmental problems worldwide. With more than 800 titles in print and some 40 new releases each year, we are the nation's leading publisher on environmental issues. We identify innovative thinkers and emerging trends in the environmental field. We work with world-renowned experts and authors to develop cross-disciplinary solutions to environmental challenges.

Island Press designs and implements coordinated book publication campaigns in order to communicate our critical messages in print, in person, and online using the latest technologies, programs, and the media. Our goal: to reach targeted audiences—scientists, policymakers, environmental advocates, the media, and concerned citizens—who can and will take action to protect the plants and animals that enrich our world, the ecosystems we need to survive, the water we drink, and the air we breathe.

Island Press gratefully acknowledges the support of its work by the Agua Fund, Inc., The Margaret A. Cargill Foundation, Betsy and Jesse Fink Foundation, The William and Flora Hewlett Foundation, The Kresge Foundation, The Forrest and Frances Lattner Foundation, The Andrew W. Mellon Foundation, The Curtis and Edith Munson Foundation, The Overbrook Foundation, The David and Lucile Packard Foundation, The Summit Foundation, Trust for Architectural Easements, The Winslow Foundation, and other generous donors.

The opinions expressed in this book are those of the author(s) and do not necessarily reflect the views of our donors.

VITAL SIGNS

Printed and bound in Great Britain by
Marston Book Services Limited, Oxfordshire

VITAL SIGNS

VOLUME 20

The Trends That Are Shaping Our Future

WORLDWATCH INSTITUTE

Michael Renner, *Project Director*

Eric Anderson   Danielle Nierenberg
Samantha Bresler   Alexander Ochs
Seyyada A. Burney   James Paul
Xing Fu-Bertaux   Grant Potter
Gary Gardner   Judith Renner
Lori Hunter   Michael Renner
Mark Konold   Laura Reynolds
Supriya Kumar   Reese Rogers
Petra Löw   Cameron Scherer
Matt Lucky   Katie Spoden
Shakuntala Makhjani   Catherine Ward
Evan Musolino

Linda Starke, *Editor*
Lyle Rosbotham, *Designer*

ISLANDPRESS

Washington | Covelo | London

+-15

Contents

Technical Note

Units of measure throughout this book are metric unless common usage dictates otherwise. Historical data series in *Vital Signs* are updated in each edition, incorporating any revisions by originating organizations. Unless noted otherwise, references to regions or groupings of countries follow definitions of the Statistics Division of the U.N. Department of Economic and Social Affairs. Data expressed in U.S. dollars have for the most part been deflated (see endnotes for specific details for each trend).

Acknowledgments

Each edition of *Vital Signs* is the product of collaboration among a diverse group of people. Worldwatch staff researchers and outside authors contribute the most visible part of this book. The contents of each article that you, the reader, see are the end result of many hours of conducting research, sifting through a broad range of source materials, crunching numbers, and providing expert analysis. The articles in this book were first released on our companion online site, at vitalsigns.world watch.org, over the course of 2012 and early 2013.

Many individual and institutional funders, as well as our exceedingly generous Board, provide the support without which our work would not be possible. For their support of not just this volume but also of our flagship publication, *State of the World*, and a range of other reports and projects, we are deeply grateful to a wide range of funders. They include the Barilla Center for Food & Nutrition; Caribbean Community; Climate and Development Knowledge Network; Compton Foundation, Inc.; The David B. Gold Foundation; Del Mar Global Trust; Elion Group; Energy and Environment Partnership with Central America; Ford Foundation and the Institute of International Education, Inc.; Green Accord International Secretariat; Hitz Foundation; Inter-American Development Bank; International Climate Initiative of the German Federal Ministry for the Environment, Nature Conservation and Nuclear Safety; International Renewable Energy Association; MAP Sustainable Energy Fellowship Program; Ministry for Foreign Affairs of Finland; Ray C. Anderson Foundation; Renewable Energy Policy Network for the 21st Century; Richard and Rhoda Goldman Fund and the Goldman Environmental Prize; Shenandoah Foundation; Small Planet Fund of RSF Social Finance; Steven C. Leuthold Family Foundation; Town Creek Foundation; Transatlantic Climate Bridge of the German Federal Ministry for the Environment, Nature Conservation and Nuclear Safety; United Nations Population Fund; Wallace Global Fund; the Victoria and Roger Sant Founders Fund of the Summit Fund of Washington; V. Kann Rasmussen Foundation; Weeden Foundation; William and Flora Hewlett Foundation; and Women Deliver, Inc.

This edition was written by a team of 23 researchers. In addition to outside contributors Lori Hunter, Petra Löw, James Paul, and Judith Renner, a group of veteran Worldwatch researchers, former colleagues, and interns put their fingers on the globe's pulse. They include Eric Anderson, Samantha Bresler, Seyyada A. Burney, Xing Fu-Bertaux, Gary Gardner, Mark Konold, Supriya Kumar, Matt Lucky, Shakuntala Makhijani, Evan Musolino, Danielle Nierenberg, Alexander Ochs, Grant Potter, Michael Renner, Laura Reynolds, Reese Rogers, Cameron Scherer, Katie Spoden, and Catherine Ward.

Vital Signs authors receive help from experts who kindly offer data and insights on the trends we follow. We give particular thanks this year to Colin Couchman at IHS Automotive, Nicola Kelly and Betsy Dribben at the International Co-operative Alliance, Lisa Stolarski at the National Cooperative Business Association, and Stephen Kretzmann at Oil Change International.

Long-time readers of this series will be familiar with one name in particular, that of editor Linda Starke, who not only watches over matters of grammar, punctuation, and style but more broadly brings consistency to the writings of a diverse set of authors. Once all the texts are edited and ready to be assembled into one volume, graphic designer Lyle Rosbotham ensures a consistent and visually pleasing layout and finds a suitable cover image for the book. And this year he redesigned the layout of the book in line with the new look of our annual *State of the World*.

During 2012 and early 2013, we bade farewell to a number of colleagues who will be sorely missed, including the head of our food and agriculture team, Danielle Nierenberg, and Bernard Pollack, who was in charge of communications. Supriya Kumar has done a superb job since taking over as Communications Manager, as did Cameron Scherer, who played a central role in running the *Vital Signs Online* site but who left Worldwatch in early 2013 in search of new adventures.

No less important are the people who work hard to ensure that our work is funded and that the office is well managed. We thank in particular veterans Barbara Fallin (Director of Finance and Administration) and Mary Redfern (Director of Institutional Relations), Development Associates Courtney Dotson and Grant Potter, and Executive Assistant Andrew Alesbury. A special thanks is also due to Robert Engelman, who has injected new vigor and vision into the Institute's work since taking over as Worldwatch President in 2011. And the chairman of our Board, Ed Groark, provides strategic vision, management expertise, and fundraising acumen.

Finally, we express gratitude to our colleagues at Island Press. Maureen Gately, Jaime Jennings, Julie Marshall, David Miller, Sharis Simonian, and Brian Weese have shared many exciting ideas about how to make *Vital Signs* an even more exciting product. We look forward to many years of a productive relationship with them.

Michael Renner
Project Director
Worldwatch Institute
1400 16th Street, N.W.
Washington DC 20016
vitalsigns.worldwatch.org

Overview:
Peak Production from a Planet in Distress

Michael Renner

The world's production and consumption trends for energy, grain, meat, fish, metals, cars, and other commodities and products continue to point upward. They were only temporarily interrupted, like a collective hiccup, by the global financial crisis. Our economic systems and theories are programmed to squeeze ever more resources from a planet in distress—whether it be more oil and gas from underground deposits, more milk from a cow, or more economic surplus from the human workforce.

Yet this apparent success weakens biodiversity and undermines the resilience of natural and human systems in the face of a changing climate, rising water scarcity, disease outbreaks, and other challenges. Attempts to adjust some of the economic signals (such as carbon pricing through cap-and-trade schemes) have not altered the fundamental dynamics. A mixture of population growth, consumerism, greed, and short-termism seems to be inexorably driving human civilization toward a showdown with the planet's limits.

A Destructive Pillar

The energy system is the core pillar of modern civilizations—and the source of the greatest threat to its continued existence, in the form of runaway greenhouse gas (GHG) emissions. Although its share as a source of primary energy worldwide has declined for a dozen consecutive years, oil remained the largest contributor at 33 percent in 2011. To satisfy the world's relentless appetite for energy, both harder-to-reach and unconventional forms of oil are increasingly exploited, including deepwater deposits, Venezuelan heavy oil, and Canadian oil sands.

Yet as climate scientist James Hansen has warned, extracting more carbon-intensive deposits will spell "game over" for efforts to avoid catastrophic climate change. This concern is intensified by the fact that the share provided by coal, the dirtiest of fuels, has climbed to 28 percent—its highest point in 40 years. Global atmospheric carbon dioxide (CO_2) concentrations reached 391.3 parts per million in 2011—which was 45 percent above 1990 and far above safe levels.

Some observers think that a transition from coal to natural gas, a less carbon-intensive fuel, can help reduce GHG emissions. But the growing exploitation of unconventional gas through "fracking" has triggered environmental and health concerns as well as worries that "cheap" gas could undermine the growth of renewable energy.

Worldwide renewable energy investments reached a new peak of $257 billion in 2011. The bulk of these investments—$147.4 billion—went to solar power, surpassing wind. For now, wind and solar power remain much smaller energy sources than hydropower. But large-scale dams such as Belo Monte, for which Brazil recently broke ground, disrupt ecosystems and displace large numbers of people. Small hydro is a better alternative. Other renewables are on the rise as well. For example, geothermal power has almost doubled in capacity since 1990, but at slightly more than 11 gigawatts, it remains comparatively small.

Additional investments are needed to facilitate the integration of renewables into aging grid infrastructures. Global spending for "smart grid" technologies rose 7 percent in 2012, totaling $13.9 billion worldwide.

Replacing fossil fuels with cleaner energy can be accelerated by changing the pattern of energy subsidies. Estimates of subsidies for fossil fuels in 2012 range from $775 billion to more than $1 trillion, compared with subsidies for renewable energy that totalled a mere $66 billion in 2010.

Carbon capture and storage (CCS) is billed as a pro-climate measure, but this may ultimately prolong the life of carbon-intensive industries. In fact, CCS is often being funded to extract additional fossil fuels—for instance, when captured CO_2 is injected into wells for enhanced oil recovery. The money needed to build enough CCS capacity to make a significant dent in emissions, up to $3 trillion by 2050, could be better spent in pursuit of clean energy and energy efficiency.

Car-centered transportation is one of the largest consumers of fossil fuels. Following a plunge in output triggered by the global economic crisis, world auto production is roaring back to new peaks—estimated at more than 80 million light vehicles in 2012. Increasing numbers and growing distances driven threaten to overwhelm recent fuel economy advances. Hybrid and electric vehicles still account for a very small share—less than 2 percent—of total production.

Record Production, Gathering Challenges

The repercussions of a warmer planet may be felt most strongly in the world's farm fields. As the experiences in many areas during 2012 showed, yields could suffer from an onslaught of extreme temperatures, floods, droughts, and other climate impacts.

Defying the challenges, global grain production was at a historic peak early in 2012. But water scarcity is a rising problem in various parts of the world. Around 1.2 billion people live in areas of physical water scarcity, while another 1.6 billion face economic water shortages, a symptom of poverty.

Farming could be made more resilient through measures like crop diversification, agroforestry, and rainwater harvesting but also by empowering women farmers. Women produce 60–80 percent of the food in developing countries but own less than 2 percent of the land.

As is the case with energy, subsidies have had a negative impact—pushing the exploitation of aquifers beyond sustainable levels in some of the major agricultural-producer countries, leading to salinization and waterlogging. More-efficient drip irrigation has the potential to reduce water use by as much as 70 percent while increasing output by 20–90 percent. Within the last two decades, efficient irrigation methods have increased more than sixfold, to over 10.3 million hectares (or 3 percent of all land equipped for irrigation).

Organic agriculture is another solution. It uses up to 50 percent less fossil fuel energy than conventional farming, helps stabilize soils and improve water retention, and allows for higher biodiversity. Since 1999, the land area farmed organically has expanded more than threefold, to 37 million hectares. But it still accounts for only about 1 percent of total agricultural land.

Although alternative techniques form an important part of more-sustainable farming, the nature of the challenge is social and economic. Land is distributed very unequally, and at least 1 billion people worldwide do not have access to sufficient food—essentially because they are poor and marginalized. The phenomenon of "land grabbing" (in which foreign investors acquire land for export production, biofuels development, and other purposes) is leading to the displacement of local farmers who have insecure land tenure. Since 2000, an estimated 70.2 million hectares of agricultural land, principally in poorer countries, have been sold or leased, equal to less than 2 percent of the world's agricultural land.

Like grain production, meat production is also at a historic peak, but drought, disease outbreaks, and rising prices of livestock feed have slowed the rate of growth. The number of farm animals worldwide has more than doubled to close to 27 million in 2010. Zoonotic diseases—transmitted between animals and humans—cause about 2.7 million human deaths each year. Many of them can be traced to the factory farms that now account for 72 percent of global poultry production, 55 percent of pork production, and 43 percent of egg production. Factory farms also contribute to GHG emissions, produce large amounts of waste, use huge quantities of water and land, and cause biodiversity loss.

One area where planetary resources are increasingly tapped out is the wild fish catch, which has stagnated during the last two decades. More than 57 percent of all fisheries are fully exploited and another 30 percent are overexploited. But fish farming has pushed total catch levels to an all-time high. This is a double-edged sword, though, given that aquaculture can contribute to habitat destruction, waste disposal, invasions of exotic species and pathogens, and depletion of wild fish stock. More-sustainable management of aquaculture is essential.

Disparities

It might seem that a growing human population has no choice but to keep raising the levels of production and consumption. Indeed, more and more people in developing countries want to copy the ways of the Western world, with its materials-intensive lifestyles, fashions, and diets. Advertising has long gone global, and although worldwide expenditures have not yet rebounded to the levels prior to the start of the economic crisis, 2012 saw a growth in ad spending of 3.3 percent (and in the Asia Pacific region, a much higher 7.9 percent) to reach $497.3 billion.

But on a per capita basis, the gaps between "industrial" and "developing" countries remain huge, whether you look at CO_2 emissions, metals use, waste generation, or many other indicators. For instance, although meat consumption is rising fast in developing countries, the average annual consumption of 32.3 kilograms is still less than half of the 78.9 kilograms per person in industrial countries.

Industrial countries have in-use stocks of aluminum of 350–500 kilograms per person, which is 10–14 times as much as in developing countries. The ratios for copper, iron, and zinc are not quite as high, but they are nonetheless pronounced.

This is the case even though global metals production has surged to unparalleled heights in the last two decades, reaching almost 1.5 billion tons in 2010.

Richer countries not only use larger quantities of metals and other materials, they also generate more waste. In the United States, each person on average is responsible for 2.58 kilograms of waste per day, compared with about 1 kilogram in China, Brazil, and Russia and just 0.34 kilograms in India.

Urban-rural differences remain pronounced. Cities are home to the most affluent people on Earth but also to vast slums that house more than 800 million people. Encouragingly, however, the sanitation and water access for 227 million people was improved between 2000 and 2010 to the point where they are no longer considered slum dwellers.

In virtually all countries, the divide between haves and have-nots is increasing. The economic crisis has deepened many economic challenges, with unemployment rising from 169 million in 2007 to about 197 million in 2012. As far back as the 1980s, wages in many countries stopped keeping pace with improvements in labor productivity. A rising share of wealth thus goes to profits, and there is a growing gap between top earners and everyone else.

The extremely unequal distribution of income and wealth that has emerged worldwide has profound consequences, determining who has an effective voice in matters of economics and politics—and thus how countries address the fundamental challenges before them.

Another kind of disparity is found in vulnerability to climate-intensified disasters, which often hit poorer countries much harder. The years since 1980 have brought an annual average of 630 disaster events, of which 86 percent were weather-related (mostly storms and floods), along with 73,000 fatalities during disasters. The severe drought in the Horn of Africa from October 2010 until September 2011 caused widespread famine and large-scale migratory movements, particularly in Somalia and Kenya, and an estimated 50,000 people lost their lives.

More-frequent and destructive disasters could have far-reaching consequences for the societies affected, and there is growing discussion about the impacts on stability. Some analysts have warned that large-scale population displacements and conflicts could occur in coming years. Environmental changes such as rainfall shortages and heat waves may interact with other pressures such as persistent poverty and population growth. Migration could be both a symptom of the repercussions and a coping strategy.

Solutions: Technical or Institutional?

There is no shortage of alternatives to change the destructive trajectory that humanity finds itself on. Renewables and efficient irrigation are two practical options among many others. Recycling is another—reducing the need for logging and mining as well as saving energy. A more circular economy that promotes recycling, reuse, and remanufacturing could address many of the challenges before us. In Japan, for example, resource productivity is on track to more than double by 2015 over 1990 levels.

But it would be a mistake to think that the solution lies in science and technology per se. Scientists have enormously improved our understanding of the climate system, for instance, yet science alone is clearly not able to drive policy in the directions that are needed. Such an outcome rests on the more difficult task of changing the dominant ethics and values and reorienting our political and economic systems toward sustainability. A broad array of efforts is needed on all levels.

On the local level, co-operatives can be part of an alternative approach to the dominant corporate model. About 1 billion people in 96 countries belong to a co-operative—among them worker, consumer, producer, and purchasing co-operatives as well as credit unions and other organizations. At the international level, multiplying global crises—including climate change, financial instability, resource limits, transboundary disease, and poverty—need commonly decided international solutions. But the entire annual budget of the United Nations system runs to only about $30 billion, less than half the size of the budget of New York—home to U.N. headquarters.

This might serve as an involuntary parable for our times. We are, in effect, shortchanging ourselves. Humanity can do far better, whether the challenge is climate stabilization, poverty eradication, social justice, or international cooperation. But we need to get serious about these tasks instead of largely consigning them to the margins. That requires political change.

Energy and Transportation Trends

Oil rig complex in the Gulf of Mexico

For additional energy and transportation trends, go to vitalsigns.worldwatch.org.

Growth in Global Oil Market Slows

Shakuntala Makhijani

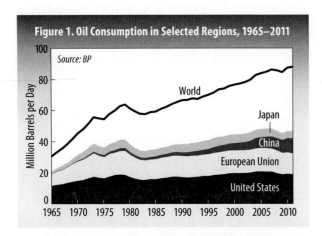

Figure 1. Oil Consumption in Selected Regions, 1965–2011

Source: BP

(Chart: Million Barrels per Day, years 1965 to 2010, showing World, Japan, China, European Union, United States)

Global oil consumption increased by 0.7 percent in 2011 to reach an all-time high of 88.03 million barrels per day.[1] (See Figure 1.) This rate of increase was considerably slower than in 2010, when oil consumption rose by 3.3 percent following a decline of 1.3 percent in 2009 due to the global financial crisis.[2] China's oil consumption increased by 5.5 percent in 2011, and China accounted for about 85 percent of global net growth.[3] An increase in oil consumption of 5.7 percent in the former Soviet Union contributed another 37 percent of net growth.[4] But these increases were offset by declines in the United States and European Union, where oil consumption fell by 1.8 and 2.8 percent.[5]

The gap in oil consumption between countries in the Organisation for Economic Co-operation and Development and all other countries narrowed further in 2011, with the two groups respectively accounting for 51.5 and 48.5 percent of total oil consumption.[6] Oil remained the largest source of primary energy worldwide in 2011, but its share fell for the twelfth consecutive year to 33 percent.[7]

To meet continued growth in demand, global oil production rose for the second year in a row, by 1.3 percent in 2011, to reach 83.58 million barrels per day.[8] (See Figure 2.) Most of this increase was driven by higher production in countries that belong to the Organization of Petroleum Exporting Countries (OPEC), which overall grew by 3 percent in 2011 due to significant production growth in Iraq, Kuwait, Qatar, Saudi Arabia, and the United Arab Emirates.[9] Meanwhile oil production in non-OPEC countries fell slightly, by 0.1 percent.[10] Oil production growth was slow compared with natural gas and coal production, which grew by 3.1 and 6.1 percent, respectively, in 2011.[11]

Political unrest in the Middle East and North Africa had a significant effect on oil production in certain countries in the region. Output in Libya fell by 71 percent in 2011—from 1.7 million barrels per day (2 percent of global production in 2010) to just 479,000 barrels (0.6 percent of global output) due to the disruptions from the civil war.[12] At the same time, tense political situations and violence in Iran, Syria, and Yemen resulted in production declines of 0.6, 13.7, and 24 percent in 2011.[13]

The growth in OPEC oil production led to a widening gap between OPEC and non-OPEC production, which respectively accounted for 43 and 41 percent

Shakuntala Makhijani is a research associate on the Worldwatch Institute Climate and Energy team.

of global production in 2011, with the remaining 16 percent produced by the former Soviet Union.[14] (See Figure 3.) Saudi Arabia increased its oil production by 12.7 percent to 11.2 million barrels per day in 2011, regaining its position as the world's largest producer and overtaking Russia, where production growth slowed to 1.2 percent.[15] Saudi Arabia's decision to increase production, in part a response to concern over the impact of the Libyan civil war on global oil markets, drew sharp criticism from other OPEC members, particularly Iran, that were concerned about the dampening effect this would have on oil prices.[16]

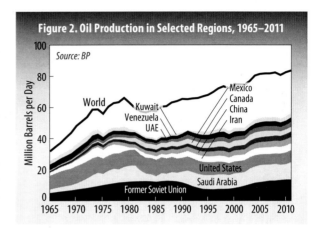

Figure 2. Oil Production in Selected Regions, 1965–2011

The impacts of the April 2010 Deepwater Horizon offshore drilling rig blowout and oil spill on U.S. offshore oil production are expected to persist for the next few years.[17] The International Energy Agency expects that by 2015, delays due to the subsequent temporary drilling moratorium will drop production from offshore drilling in the Gulf of Mexico 300,000 barrels a day (about 3.8 percent of 2011 U.S. oil production) lower than previously projected.[18] Since the incident, the United States has implemented only a few new rules on standards for oil-well permits, blowout preparedness, inspections, and workplace safety, which are not expected to significantly affect future levels of offshore oil drilling.[19] Although many key in-

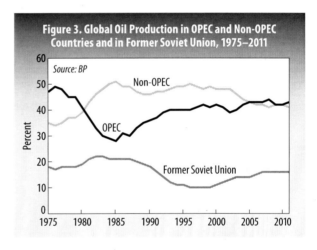

Figure 3. Global Oil Production in OPEC and Non-OPEC Countries and in Former Soviet Union, 1975–2011

vestigations and lawsuits are ongoing, political pressure due to continuing economic difficulties and high oil prices has led to a renewed push for offshore oil drilling in the Gulf of Mexico as well as Alaska.[20]

The global impacts of the April 2010 accident have been limited thus far, with reviews in most countries finding that existing safety requirements suffice to prevent similar accidents.[21] Despite expanding offshore drilling efforts, the share of offshore oil is expected to remain steady at 30 percent of global oil production due to declining output from North Sea and Mexican offshore oil wells.[22] Deepwater oil production is expected to constitute a growing portion of this production and is projected to go from 6 percent of total global oil supply today to 9 percent by 2016.[23]

Oil prices reached all-time highs in mid-2008 but then fell sharply as the global financial crisis drove demand down.[24] (See Figure 4.) With OPEC's decision to cut production targets in the first quarter of 2009, world crude prices began to recover, and average annual prices for West Texas Intermediate crude reached $94.83 per barrel in 2011, close to the average 2008 price of $99.67 per barrel.[25]

Against the backdrop of fluctuating oil prices and concerns about supply risk,

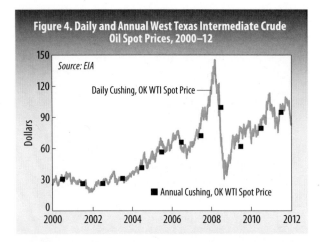

Figure 4. Daily and Annual West Texas Intermediate Crude Oil Spot Prices, 2000–12

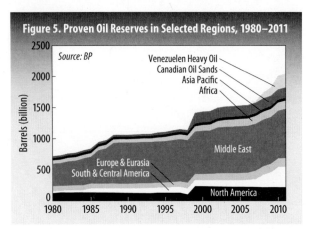

Figure 5. Proven Oil Reserves in Selected Regions, 1980–2011

many countries are paying more attention to their dependence on imports and the stability of the countries they purchase oil from. In 2011, the United States imported 60 percent of the oil it needed, and Europe imported 90 percent.[26] Imports accounted for 68 percent of China's oil consumption, while Japan actually imported slightly more oil than it consumed in 2011.[27]

The Middle East remains the world's largest oil exporter, accounting for 36.2 percent of exports in 2011 and a growing share of the global market.[28] Russia and the Asia Pacific region were the second and third largest exporters, with 15.9 and 11.4 percent, respectively.[29] Oil exports from North Africa fell by 32.8 percent in 2011 due largely to the disruptions in oil production caused by political instability in the region.[30] Exports from the United States grew by 19.4 percent in 2011, faster than in any other region, but they accounted for only 4.7 percent of the global market.[31]

Meanwhile, global proven oil reserves, including natural gas condensate and natural gas liquids in addition to crude oil, have been increasing since 1980. They grew by 1.9 percent in 2011 to reach an estimated 1,652.6 billion barrels (1,821.8 billion barrels if Canadian oil sands are included and 2,041.8 billion barrels with Venezuelan heavy oil).[32] (See Figure 5.) OPEC countries control 72.4 percent of global oil reserves, and the Middle East has the largest share of reserves of any region, at 48.1 percent of the total.[33] Venezuela has the largest share of crude oil reserves of any country, with 296.5 billion barrels (17.9 percent of the global total).[34] Heavy oil (which is not typically included in global oil reserve estimates) in Venezuela's Orinoco belt adds another 220 billion barrels to that country's reserves.[35] Saudi Arabia has the second largest share of any country, with 16.1 percent of global oil reserves.[36]

Canadian oil sands proven reserves remained steady between 2010 and 2011, at 169.2 billion barrels, or 9.3 percent of global oil reserves when oil sands are included.[37] Canada's oil sand reserves currently under development likewise remained steady between 2010 and 2011, at 25.9 billion barrels.[38]

Canadian oil sands reserves became a high-profile issue in 2011 thanks to protests by environmental groups over the proposed Keystone XL pipeline that would bring oil sands from Alberta in Canada to Texas refineries on the Gulf Coast.[39] Faced with a short deadline imposed by Republicans in Congress in January 2012, President Obama rejected the original proposed pipeline route, citing the risk to groundwater resources in the Ogallala aquifer and the ecologically sensitive Sand-hills region of Nebraska.[40] TransCanada, the company seeking approval for the pipeline, submitted a new application to the State Department in May 2012 with an updated route that bypasses the Sandhills.[41] A decision on this new proposal was not expected until early 2013.[42] Proponents of the oil pipeline claim benefits of increasing oil production from a geopolitically stable country, job creation, and the need for new oil sources in a tightening global oil market.[43] Environmental groups continue to oppose the revised route due to unresolved concerns about developing Alberta oil sands, including the climate impacts of tapping this energy-intensive oil source, the high water requirements for oil sands development, the risk of oil spills along the pipeline, and landscape alteration and toxic waste streams from oil sands mining.[44] Despite these concerns, oil sands account for about half of Canada's crude oil production, a share that is expected to rise in the future.[45]

Global Coal and Natural Gas Consumption Continue to Grow

Matt Lucky and Reese Rogers

Global consumption of coal and natural gas continued to grow in 2011. Coal use increased by 5.4 percent to 3,724.3 million tons of oil equivalent (mtoe) from the end of 2010 to the end of 2011.[1] Demand for natural gas grew by 2.2 percent in 2011, reaching 2,905.6 mtoe.[2]

Although oil remains the world's leading energy source, coal and natural gas continue to grow in importance. Both are the primary fuels for the world's electricity market. And because they often act as substitutes for each other, their trends need to be looked at together.

Spurred mainly by demand growth in China and India, coal's share in the global primary energy mix reached 28 percent in 2011—its highest point since the International Energy Agency began keeping statistics in 1971.[3] While the United States remained one of the world's largest coal users, consumption growth in 2011 was concentrated among countries that are not part of the Organisation for Economic Co-operation and Development (OECD), including China and India.[4] (See Figure 1.) Consumption in non-OECD countries grew by 8.4 percent to 2,625.7 mtoe.[5] These countries accounted for 70.5 percent of global coal consumption in 2011.[6] The bulk of this usage occurs in the electricity-generating sector, with smaller amounts used in steelmaking.[7]

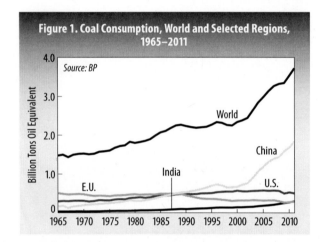

Figure 1. Coal Consumption, World and Selected Regions, 1965–2011

Source: BP

China remains the largest coal consumer in the world, accounting for 49.4 percent of global use in 2011.[8] The country maintained the rapid rate of coal demand growth seen over the last decade, with consumption growing 9.7 percent to 1,839.4 mtoe.[9] Over the period 2001–11, China accounted for 80 percent of global coal demand growth.[10] Much of this coal is used in the domestic power sector. In 2010, almost 80 percent of China's power generation came from coal-fired units, and in 2011 China overtook the United States as the largest power producer in the world.[11]

India also figures prominently in the growth of the international coal market as the second largest contributor to demand growth. India's coal consumption grew 9.2 percent to 295.6 mtoe in 2011.[12] It remains the third largest consumer of coal in the world, after surpassing the European Union in 2009.[13]

Coal demand in the United States, the second largest coal user, decreased by 4.5

Matt Lucky is a Climate and Energy research associate and **Reese Rogers** is a MAP Sustainable Energy Fellow at Worldwatch Institute.

percent in 2011 and continued to fall in 2012 due to the shale gas boom and the abundance of cheap natural gas.[14] Even with the decrease in demand, the United States still accounts for 45 percent of coal demand within the OECD.[15] Over 90 percent of coal consumption in the United States occurs in the power sector.[16] As of August 2012, net generation from coal accounted for 38.5 percent of U.S. electricity output, a rapid fall from 45 percent in 2010.[17]

Global coal production increased by 6.1 percent to 3,955.5 mtoe (6,941 million tons of coal) in 2011, making it the fastest-growing fossil fuel.[18] Coal production, like coal consumption, is mainly concentrated in China. (See Figure 2.) While China is far and away the largest producer, it does not hold the largest proven reserves in the world. That title belongs to the United States, with 28 percent of the global total.[19] China holds 13 percent, and Russia, Australia, and India account for 18, 9, and 7 percent, respectively.[20] Thus these five countries accounted for three quarters of the proven reserves in the world as well as three quarters of global production in 2011.[21]

China alone accounted for 46 percent of global coal production in 2011, with an output of 1,956 mtoe.[22] With proven reserves at 114,500 million tons, China's current levels of production could continue for over three decades.[23] But demand and production are likely to continue rising during that time, a trend reflected in recent changes to China's twelfth Five-Year Plan (2011–15), which seeks to expand domestic coal mining capacity to 4.1 billion tons.[24]

In 2011, the United States produced 556.8 mtoe, compared with its peak at 596.7 mtoe in 2008.[25] While domestic demand for coal has declined, coal exports are increasing, with net exports in 2011 reaching nearly 55.76 mtoe.[26] Data through August 2012 show that U.S. coal exports are growing at a rate not seen since the 1979–81 export boom and that 2012 export numbers should more than double the level of 2009.[27] While demand for American coal is growing in Asia, the United States still exports more coal to Europe than to the entire rest of the world—sales that have been bolstered by growing exports of steam coal.[28]

Coal prices increased across all major markets in 2011. (See Figure 3.) After the record high prices in 2008 and the subsequent crash, coal prices rose from early 2009 through mid-2011.[29] However, diverging trends in regional markets have created disparities in coal pricing around the world, as growing demand in Asia led to higher prices in the Pacific Basin market, while increasing exports from

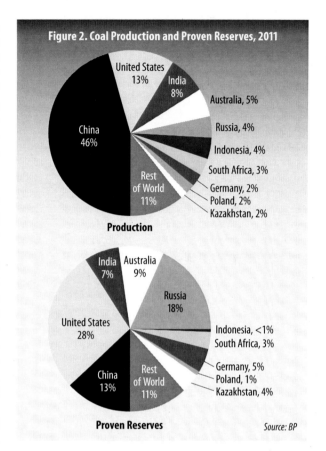

Figure 2. Coal Production and Proven Reserves, 2011

Production

United States 13%
India 8%
Australia, 5%
Russia, 4%
Indonesia, 4%
South Africa, 3%
Germany, 2%
Poland, 2%
Kazakhstan, 2%
China 46%
Rest of World 11%

Proven Reserves

India 7%
Australia 9%
Russia 18%
United States 28%
Indonesia, <1%
South Africa, 3%
Germany, 5%
Poland, 1%
Kazakhstan, 4%
China 13%
Rest of World 11%

Source: BP

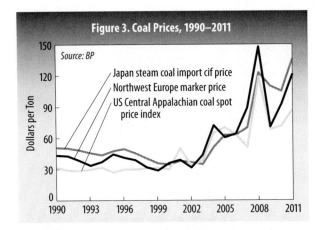

Figure 3. Coal Prices, 1990–2011

Source: BP

Japan steam coal import cif price
Northwest Europe marker price
US Central Appalachian coal spot price index

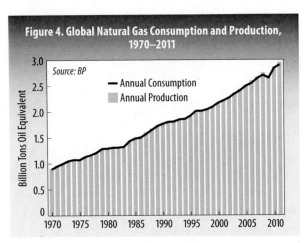

Figure 4. Global Natural Gas Consumption and Production, 1970–2011

Source: BP

— Annual Consumption
▨ Annual Production

the United States kept prices lower in Europe, where overall demand was weak.[30]

China's dominance of the global coal market places it in a crucial position of affecting import prices for the rest of the world. China's domestic price for coal sets the price for countries exporting coal to the Chinese market and then acts as a price marker for the international market.[31]

The story with natural gas is a bit different. Global consumption grew at a slower rate than coal in 2011—2.2 percent, to reach 2,905.6 mtoe.[32] (See Figure 4.) Usage grew in all regions but Europe; in fact, the European Union experienced a 9.9 percent decline in natural gas consumption—the largest on record and mainly due to a struggling economy and high natural gas prices.[33]

Natural gas accounted for 23.7 percent of global primary energy consumption in 2011, down slightly from 23.8 percent in 2010.[34] Consumption increased most significantly in East Asia, with China (21.5 percent) and Japan (11.6 percent) accounting for most of this growth.[35]

Natural gas production increased at a higher rate (3.1 percent) than consumption in 2011, reaching 2,954.8 mtoe.[36] The United States (20.0 percent) and Russia (18.5 percent) accounted for nearly 40 percent of the world's output in 2011, while Canada, Iran, and Qatar were the next largest producers at 4–5 percent each.[37] Yemen (51.3 percent), Iraq (42.0 percent), Turkmenistan (40.6 percent), and Qatar (25.8 percent) experienced the greatest production growth rates.[38] China (8.1 percent) and the United States (7.7 percent) also increased production substantially in 2011.[39]

Estimates of proven natural gas reserves increased by 6.3 percent to 187,900 mtoe, or 63.6 years at current production levels.[40] The Middle East (38.4 percent) and the former Soviet Union (36.0 percent) have the highest concentration of reserves.[41] Proven reserves increased in Turkmenistan by 9,854 mtoe alone, accounting for 89 percent of the net increase in global proven reserves.[42]

Estimates for technically recoverable natural gas resources reached 712,200 mtoe, or about 240 years at current production levels.[43] Of this total, about 41 percent are unconventional resources, including shale gas, tight gas, and coalbed methane.[44]

Most natural gas is used for power generation and in the commercial, industrial, and residential sectors. Although usage in the transportation sector remains small, an estimated 15 million natural gas vehicles are in use worldwide.[45] Most of these vehicles are found in Argentina, Brazil, India, Iran, and Pakistan, and the fleet's annual growth rate is about 4.7 percent.[46] The main barrier to growth of this sector in other regions of the world remains considerable associated infrastructural costs.[47]

The global natural gas picture has changed significantly over the past several years due to the increased production of shale gas, and nowhere is this more visible than in the United States. Natural gas consumption for power generation there increased by 27 percent from 2011 to 2012, with the gas largely used to replace coal.[48] Over the same period of time, coal consumption for power generation declined by 20 percent.[49] Because natural gas emits about half as many greenhouse gases (GHG) per unit of electricity produced as coal, this shift—combined with energy efficiency gains and increased renewable energy production—helped the United States reduce its 2012 GHG emissions from energy consumption by about 15 percent relative to the peak level in 2007.[50] It is estimated that by 2035 natural gas will surpass oil as the most important fuel in the U.S. energy mix.[51]

Natural gas prices increased significantly in 2011 in all regions but North America.[52] (See Figure 5.) Japanese liquefied natural gas (LNG) import prices increased by 35 percent in 2011.[53] Despite these high prices, Japan continues to import more LNG to make up for nuclear shutdowns following the disaster at the Fukushima plant. On the other hand, natural gas prices reached as low as $2.10 per million Btu in the United States in June 2012, highlighting the extremity of regional price differences for natural gas.[54]

Nevertheless, a global price for natural gas is more likely to occur in the future due to increasing short-term markets and greater trade flexibility.[55] The construction of liquefaction plants in the United States would help to connect the U.S.-Asia markets and bring about a more global natural gas market.[56]

LNG is 600 times denser than naturally occurring gas and therefore more economical to ship. LNG markets continued to grow in 2011. The share of natural gas trade represented by LNG reached 32.2 percent in 2011, up from 30 percent in 2010.[57] LNG trade increased by 10 percent in 2011, while pipeline trade increased by only 1.3 percent over the same period.[58]

LNG imports were concentrated in East Asia in 2011 (see Figure 6), with China, Japan, South Korea, and Taiwan accounting for more than half of global LNG imports.[59] For the fourth year in a row, U.S. LNG imports declined, reflecting the

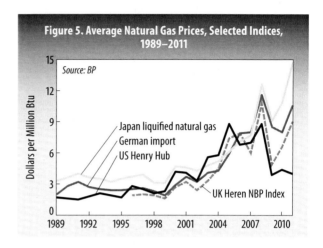

Figure 5. Average Natural Gas Prices, Selected Indices, 1989–2011

Source: BP

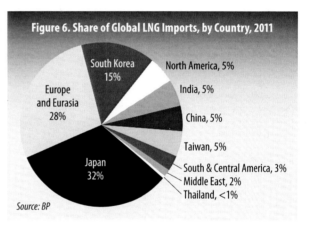

Figure 6. Share of Global LNG Imports, by Country, 2011

Source: BP

abundance and low cost of domestic natural gas resources.[60] Interestingly, despite the European Union's 9.9 percent reduction in natural gas consumption, its LNG imports increased by 5 percent in 2011.[61]

By the end of 2011, the world's installed LNG export capacity reached 337 mtoe.[62] This figure is expected to grow, however, with 97 mtoe currently under construction worldwide (mostly in Australia).[63] LNG export capacity has the potential to grow even more substantially over the next decade, but much of this depends on whether the United States becomes an LNG exporter in the future.

Continuation of these growth trends in the global coal and natural gas sectors depends on numerous factors, and while continued growth is likely, it is still uncertain. Policies to reduce the environmental and health effects of coal combustion and the development of new technologies in the power sector could contribute to stagnating demand for coal. At the same time, increasing global concern about GHG emissions and climate change could lead to a greater transition from coal to natural gas, although environmental concerns about fracking and the possibility that cheap gas might undermine growth in renewables could change this. Nevertheless, the rapid development of China and India, which simply need fuel to maintain their economic growth, will likely maintain the consumption growth rates of these two fossil fuels.

China Drives Global Wind Growth

Mark Konold and Samantha Bresler

In 2011, global wind power capacity topped out at 238,000 megawatts (MW) after adding just over 41,000 MW.[1] (See Figure 1.) This means that the global capacity grew by 21 percent in 2011—lower than the 2010 rate of 24 percent and markedly lower than the 2009 rate of 31 percent.[2] Nonetheless, the world now has four times as much installed wind power capacity as in 2005, just seven years ago.[3]

China led the way with 43 percent of global capacity additions during 2011, followed by the United States with 17 percent, India with almost 7 percent, and Germany with 5 percent.[4] In terms of cumulative capacity, China has a commanding 26 percent of global installed capacity.[5] (See Figure 2.) It is followed by the United States, Germany, Spain, and India. A total of almost $75 billion was invested in wind energy installations in 2011, which was 22 percent less than invested in 2010.[6]

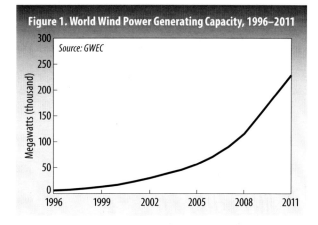

Figure 1. World Wind Power Generating Capacity, 1996–2011

Source: GWEC

For the second year in a row, China set the pace and propped up the industry, increasing its total capacity by 40 percent over 2010 levels.[7] China added just over 17,000 MW of new capacity, bringing its grand total to nearly 63,000 MW.[8] There remains an important gap between total installed capacity and actual electricity available for use from wind power, however. Despite having the most installed wind capacity, China still struggles to use all the electricity its turbines generate.

In 2011, China's cumulative wind capacity generated 69 terawatt-hours (TWh) of electricity, 1.5 percent of the country's total supply.[9] But just under 17 percent of that electricity never made it to the grid.[10] In fact, the provinces of Inner Mongolia and Gansu lost 23 and 25 percent of their generated capacity due to technical difficulties.[11] China plans to have an electrical grid strong enough to fully integrate its total installed capacity by 2015.[12] During the next five years, the State Grid Corporation of China plans to invest over $400 billion in power grid construction.[13] At the end of 2011, some 238 Smart Grid pilot projects had been implemented, with several addressing the lack of connection to wind power plants.[14]

In 2011 the United States added 6,800 MW of new capacity, bringing its total capacity to 46,919 MW.[15] Texas remained the country's leading state, with 10,377 MW of total capacity—up from 10,085 MW in 2010.[16] California and Illinois have the second and third largest capacities, with 921 MW and 693 MW, respectively.[17]

Mark Konold is the Caribbean program manager for Worldwatch Institute's Climate and Energy Program. **Samantha Bresler** was a Climate and Energy research intern at Worldwatch.

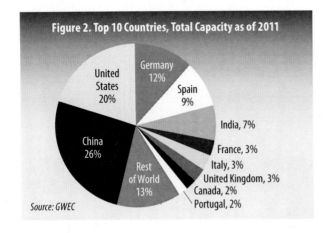

Figure 2. Top 10 Countries, Total Capacity as of 2011

Germany 12%
United States 20%
Spain 9%
India, 7%
China 26%
France, 3%
Italy, 3%
Rest of World 13%
United Kingdom, 3%
Canada, 2%
Portugal, 2%

Source: GWEC

The United States generated just under 120 TWh of electricity from wind power in 2011, a 27 percent increase from 2010.[18] However, this electricity accounted for less than 3 percent of total U.S. power generation in 2011.[19]

Wind power growth in the United States owes much to the federal government's Production Tax Credit (PTC), which helped finance approximately 4,000 MW of new capacity by reducing corporate income tax by 2.2¢ for every kilowatt-hour produced.[20] Prior to the PTC's extension at the end of 2012, fears that it would lapse led to predictions that 37,000 jobs in the industry—out of a total of 75,000 currently—could be lost.[21]

The 27-member European Union (EU-27) installed 9,616 MW of wind power capacity in 2011, almost the same as in 2010.[22] Germany regained its lead position for installed wind power capacity there by adding 2,086 MW to its energy portfolio, reaching a total of 29,060 MW.[23] In 2011, wind-generated electricity provided 48 TWh of electricity, 7.8 percent of the country's electricity consumption.[24] The United Kingdom was responsible for 13 percent of the EU-27's newly installed wind capacity last year, adding 1,293 MW.[25] Rounding out the top three was Spain, which had a slow year by its standards, adding only 1,050 MW compared with more than 3,500 MW added four years ago.[26] But Spain still has the second-largest installed wind power capacity in Europe, with a total of 21,674 MW.[27] Wind power provided Spain with 42 TWh of electricity, an impressive 15.7 percent of the country's total electricity consumption.[28]

Europe's sovereign debt crisis is pushing future growth projections of wind installation down and potentially affecting investment incentives.[29] The European Union continues to scale down its use of fuel oil and nuclear power and so must find another energy source to fill the impending shortfall. It is interesting to note that while growth in wind capacity remained constant, more coal power was installed than was decommissioned.[30]

India added 3,019 MW of new wind power, bringing its total to 16,084 MW by the end of 2011.[31] India has instituted a generation-based incentive (GBI) that was set to expire in 2012. By paying for capacity, the GBI allows 80 percent of a project's investment costs to be offset in the first year of operations and provides a tax exemption for earnings for 10 years.[32] According to a new study by Lawrence Berkeley National Laboratory, India has 20–30 times more onshore wind energy potential than previously estimated.[33]

Other parts of the world showed very modest growth in wind power installations. Brazil and Mexico led the way in Latin America with 581 and 354 MW, respectively.[34] Meanwhile, Africa and the Middle East combined had a total of just 31 MW of wind capacity installed in 2011.[35]

Wind installations also expanded offshore, particularly in Europe. In 2011, the EU saw the successful installation of 866 MW, raising its total offshore capacity to 3,810 MW.[36] As a percentage of total wind installations in 2011, however, offshore accounted for just 9 percent in 2011, compared with 9.2 percent in 2010.[37]

The United Kingdom remains the region's powerhouse, with over 2,000 MW of offshore capacity online.[38] (See Table 1.) Some 58 percent of the total capacity addition in the United Kingdom in 2011 was offshore.[39] Germany, which only has 200 MW of offshore capacity online right now, recently expanded its target to 25 gigawatts (GW) of offshore capacity by 2030.[40] Denmark and Norway are also stepping up their offshore portfolios. With 868 MW of offshore capacity installed, Denmark recently raised its wind target as a percentage of its total future electricity mix from 44 to 52 percent by 2020.[41] And in Norway, the offshore Havsul wind farm is expected to bring 350 MW of capacity online, the first step in realizing the country's plan for 11 GW total.[42]

In early 2012, China's largest offshore wind project, Rudong, was connected to the grid, adding 99.3 MW to the 32 MW from offshore that was already online.[43] Rudong joins China's other offshore wind farm, Donghai Bridge, which has a capacity of 102 MW.[44] The United States continues to lag behind Europe and China in offshore wind installations. The U.S. Department of Energy is to make $180 million available over the next six years to support up to four innovative wind farms off the coast of the country or in the Great Lakes.[45] In 2011, the United States outlined a plan to achieve 54 GW of offshore wind deployment at a cost of 7–9¢ per kilowatt-hour by 2030, with an interim target of 10 GW at 13¢ per kilowatt-hour by 2020.[46]

In sales, Vestas remains the world leader, with close to 13 percent of the world market, although that share is dwindling.[47] (See Figure 3.) In an effort to challenge Siemens for offshore market share supremacy, Vestas launched its V164 last year, a 7 MW turbine designed solely for offshore wind.[48] Chinese manufacturer Sinovel dropped from second place to seventh and was replaced in the number two spot by another Chinese manufacturer, Goldwind.[49] American producer GE remained in third place but saw a drop from last year's 9.6 percent of the market.[50]

There appears to be a tendency toward larger-sized individual wind projects, both on and offshore, when considering additional infrastructure costs such as grid connection, substations, and permits.[51] In the first half of 2011, prices fell to $1.2 million per MW mainly because of supply chain efficiency improvements and

Table 1. Offshore Plans in Selected EU Countries

Country	2011 Online	2020 Planned
	(megawatts)	
United Kingdom	2,094	42,114
Denmark	857	1,200
Netherlands	247	3,953
Germany	200	21,493
Norway	2	11,042
Spain	0	6,804
France	0	6,000

Source: EWEA, Wind in Power: 2011 European Statistics (Brussels: 2012).

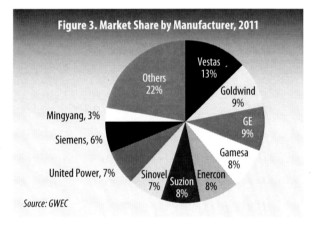

Figure 3. Market Share by Manufacturer, 2011

Vestas 13%
Goldwind 9%
GE 9%
Gamesa 8%
Enercon 8%
Suzlon 8%
Sinovel 7%
United Power, 7%
Siemens, 6%
Mingyang, 3%
Others 22%

Source: GWEC

economies of scale.[52] Competition from Chinese manufacturers and their excess capacity to build machines and flood the market also played a role.[53] In addition, the capacity factor of wind turbines (the percentage of actual output to nameplate capacity) continues to rise as better technology continues to come on to the market, further driving down turbine cost. The combination of these factors is expected to bring down the cost of wind energy by 12 percent by 2016 and to make onshore wind power truly cost-competitive with coal, gas, and nuclear power.[54]

Hydropower and Geothermal Growth Slows

Evan Musolino

Although hydropower and geothermal power are in very different stages of development, the market for these forms of electricity generation is increasing. These two sources are not subject to the variability that plagues wind and solar energy. Their greater reliability can thus be harnessed to provide baseload power.[1]

Hydropower is the older and more mature of the two technologies. Global use and installed capacity of hydropower continued to increase in 2011, reaching 3,498 terawatt-hours (TWh) and 970 gigawatts (GW) respectively.[2] Total consumption has now increased each year between 2003 and 2011.[3] (See Figure 1.) But in 2011, the growth rate slowed, registering only a 1.6 percent increase from the previous year.[4]

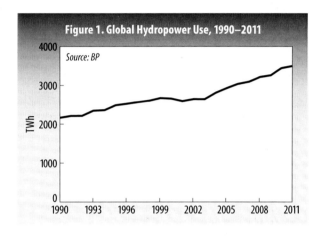

Figure 1. Global Hydropower Use, 1990–2011

Source: BP

Among members of the Organisation for Economic Co-operation and Development, hydroelectricity accounted for 5.7 percent of primary energy consumption.[5] It played a more important role in other countries, at 7.1 percent of usage, and these nations accounted for 60.2 percent of worldwide hydroelectricity consumption.[6] On a regional basis, South America and Central America are most dependent on hydroelectricity relative to total energy use. (See Table 1.)[7]

While hydropower plays its least important role in the Middle East, 2011 saw that region experience the greatest growth in hydroelectricity consumption: 21.9 percent.[8] North America was next, with a 13.9 percent increase.[9] In contrast, usage fell by 8.8 percent in Europe and Eurasia and by 0.6 percent in the Asia Pacific region.[10]

Although hydropower is produced in about 150 countries, capacity is concentrated in just a few nations.[11] China, Brazil, the United States, Canada, and Russia together accounted for 496 GW (51 percent) of global hydroelectric capacity installed at the end of 2011.[12] China continues to be the leader, with 212 GW installed, followed by Brazil (82.2 GW), the United States (79 GW), Canada (76.4 GW), and Russia (46 GW).[13]

A total of 25 GW of capacity was added in 2011, less than in previous years.[14] China, Vietnam, Brazil, India, and Canada were responsible for 75 percent of the added capacity.[15] China's 12.25 GW represented 49 percent of the worldwide gain in 2011; the country's twelfth Five-Year Plan calls for 284 GW of hydropower

Evan Musolino is a research associate in the Climate and Energy Program at Worldwatch Institute.

Table 1. Hydropower as a Share of Total Primary Energy Use, by Region, 2011

Region	Hydropower Share
	(percent)
South America and Central America	26.2
Africa	6.1
Europe and Eurasia	6.1
North America	6.0
Asia Pacific	5.2
Middle East	0.67
World	6.4

Source: Calculated from BP, Statistical Review of World Energy (London: June 2011).

capacity to be installed by 2015.[16] The four other countries expanded capacity more modestly, adding between 1.3 and 1.9 GW.[17]

Despite the potential for inexpensive, low-emission power generation from hydropower, there are significant negative consequences associated with the development of large projects. The damming of rivers to create the reservoirs needed for large-scale power generation is severely disruptive to ecosystems and can harm both animal and human populations. And building hydropower plants has its own significant emissions impacts, including from the creation of reservoirs and the large amounts of concrete needed for construction. In many cases, hydropower projects have led to the displacement of local populations and the adverse altering of downstream conditions.[18] Despite these concerns, a number of new developments occurred in 2012, as Brazil broke ground on the 11.2 GW Belo Monte hydroelectric complex.[19] In China, the final turbine in the 22.5 GW Three Gorges dam project came online in 2012, making the plant fully operational.[20]

Small hydropower (SHP) installations are becoming more popular. While it varies considerably by country, small hydropower is generally defined as a plant with an installed generating capacity below 10 megawatts (MW).[21] These installations can play a critical role in expanding energy access, as they offer the lowest life-cycle costs of all off-grid renewable technologies.[22] The potential for SHP development has been estimated at 200 GW by 2020. As of 2009, about 60 GW was installed.[23] Developing regions accounted for roughly three quarters of existing capacity. Asia has 68 percent of global SHP capacity, South America has 3 percent, and Africa has 0.5 percent.[24] Small hydropower attracted total capital investments of $5.8 billion in 2011, which was 59 percent greater than in 2010.[25] And the sector accounted for an estimated 40,000 jobs worldwide in 2011.[26]

Pumped storage hydropower—using hydropower as an energy storage solution—is increasingly gaining traction worldwide. This involves pumping water uphill into a reservoir, to be released later as needed. Pump storage facilities account for approximately 99 percent of all energy storage capacity worldwide.[27] Between 2 and 3 GW of new pumped storage hydropower capacity was added in 2011, resulting in 130–140 GW of operating capacity by the end of 2011.[28] Europe had a capacity of 45 GW as of early 2011, from 170 stations.[29] An estimated 60 plants totaling 27 GW will be added there by 2020.[30] Among individual countries, Japan (25.8 GW), the United States (22 GW), and China (18.4 GW) have most of the world's pumped storage hydropower capacity.[31]

Hydropower continues to be one of the most cost-effective renewable energy generation sources. Typical costs are 2–13¢ per kilowatt-hour (kWh) for existing grid-connected hydropower plants and 5–10¢ per kWh for new plants.[32] Micro-hydropower installations (0.1 kW to 1 MW), which are generally found in rural off-grid environments, generate at 5–40¢ per kWh.[33]

Like hydropower, geothermal resources are highly location-specific. Many countries with strong hydropower potential, including much of Latin America, the Caribbean, and Southeast Asia, have equally impressive geothermal potential. These resources have been exploited for power generation for over a century, with significant capacity being developed since the mid-1900s.[34] (See Figure 2.) However, cumulative capacity still lags far behind other renewable technologies, with 11.2 GW installed worldwide by the end of 2011.[35] Overall capacity continued to increase in 2011 despite the rate of new additions slowing to below 5 percent.[36]

Some 136 MW of new geothermal power capacity was installed in 2011.[37] The vast majority of new additions came from two major projects: a 90 MW facility in Iceland and a 42 MW plant in Costa Rica.[38] Geothermal power capacity now exists in 24 countries worldwide (with no new countries adding capacity in 2011).[39] Most of this capacity, however, is located in only a handful of nations. The United States continued to lead, with 3.1 GW installed by the end of 2011.[40] It is followed by the Philippines (1.9 GW), Indonesia (1.2 GW), Mexico (1 GW), and Italy (0.8 GW).[41] As of May 2012, New Zealand (768 MW), Iceland (661 MW), and Japan (535 MW) were the only other countries with more than 500 MW installed.[42] (See Figure 3.) Although small in terms of current capacity installed, East Africa has strong geothermal resources—with significant new developments planned in Ethiopia, Kenya, and Tanzania.[43] Indeed, new geothermal power projects are under development or consideration in some 70 countries.[44]

Worldwide, 69 TWh of geothermal electricity was generated in 2011, up from 67.2 TWh in 2010.[45] (See Figure 4.) This accounted for just 0.3 percent of global electricity generation.[46] Iceland is the global leader in geothermal power generation per person, and the nation gets 26 percent of its electricity from geothermal sources.[47] Geothermal power plants currently operate with an average capacity factor of 73 percent,

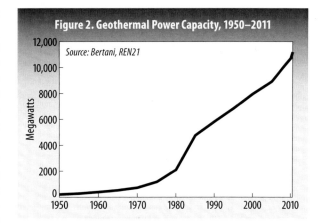

Figure 2. Geothermal Power Capacity, 1950–2011

Source: Bertani, REN21

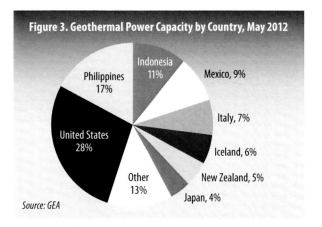

Figure 3. Geothermal Power Capacity by Country, May 2012

Indonesia 11%
Philippines 17%
Mexico, 9%
Italy, 7%
Iceland, 6%
United States 28%
New Zealand, 5%
Other 13%
Japan, 4%

Source: GEA

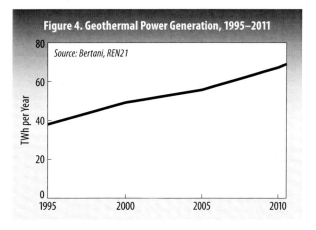

Figure 4. Geothermal Power Generation, 1995–2011

Source: Bertani, REN21

higher than any other renewable technology and rivaling fossil fuel and nuclear generation.[48] In certain cases, these plants can operate at capacity factors at or above 90 percent.[49]

The costs associated with geothermal power also closely mirror those of hydropower. Varying by geothermal technology, generation costs 5.7–10.7¢ per kWh.[50] High capital costs, associated primarily with the cost of drilling geothermal wells and the long lead time for project development, continue to challenge project developers.

Smart Grid and Energy Storage
Installations Rising

Reese Rogers

Driven by increasing shares of renewable energy in the electricity generation mix and by the need to update aging grid infrastructure, global investment in "smart grid" technologies rose 7 percent in 2012, totaling $13.9 billion worldwide.[1] A smart grid is an electricity network that uses digital information and communications technologies to improve the efficiency and reliability of electricity transport.[2] The increasing use of highly variable energy resources requires sophisticated control systems to facilitate their integration into the electricity grid.

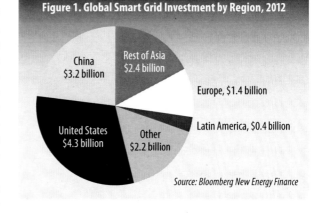

Figure 1. Global Smart Grid Investment by Region, 2012

China $3.2 billion

Rest of Asia $2.4 billion

Europe, $1.4 billion

Latin America, $0.4 billion

Other $2.2 billion

United States $4.3 billion

Source: Bloomberg New Energy Finance

The United States topped other countries in investment in smart grids, spending $4.3 billion in 2012, although that was 19 percent below the 2011 figure of $5.1 billion.[3] (See Figure 1.) China invested $3.2 billion in 2012, an increase of 14 percent over 2011.[4] Smart grid directives in the European Union (EU) drove a 27 percent increase in European spending to $1.4 billion in 2012, up from $1.1 billion in 2011.[5] Latin American investment in smart grid technology remains relatively small, totaling $400 million in 2012.[6] In addition to investments, many countries have formal nationwide development plans and regulatory frameworks for smart grids.

While the United States maintained its position as a leader in smart grids, the decline in U.S. investments in 2012 was due in part to the expiration of federal funding programs initiated under the American Recovery and Reinvestment Act in 2009.[7] Of the $3.4 billion in the federal Smart Grid Investment Grant program, about $2.3 billion had been spent as of March 31, 2012.[8] These funds have supported 99 smart grid deployment projects across the United States.[9] At the start of 2012, utilities had installed 37 million smart meters, covering 33 percent of American households.[10] Smart meters are just one of the many technologies involved in smart grid infrastructure. These electronic measurement devices gather data on energy usage and provide two-way communication with the utility for efficiency and accurate billing purposes, enabling regulatory mechanisms such as time-of-use pricing to be introduced.[11] The aggregate of utility plans to install smart meters across the country should result in 65 million units installed, covering 57 percent of American households, by 2016.[12] (See Figure 2.)

China's rising investment in smart grid technologies stems from its nationwide

Reese Rogers is a MAP Sustainable Energy Fellow with the Climate and Energy Program at Worldwatch Institute.

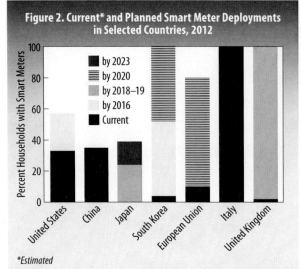

Figure 2. Current* and Planned Smart Meter Deployments in Selected Countries, 2012

Percent Households with Smart Meters

- by 2023
- by 2020
- by 2018–19
- by 2016
- Current

United States, China, Japan, South Korea, European Union, Italy, United Kingdom

*Estimated

Source: EIA, IEE, Pike Research, European Energy Review, Bloomberg, GSGF, McKinsey, Energy Saving Trust

plans to update its poorly designed and inefficient transmission system, and China is poised to surpass the United States in smart grid investment in 2013.[13] The State Grid Corporation of China has a three-phase plan to invest $601 billion in transmission infrastructure through 2020, with $101 billion slated for smart grid technology.[14] Phase 1 was completed in 2010 and included smart grid planning and pilot projects.[15] Phase 2 is expected to run from 2011 to 2015 and involves full construction and deployment of smart grid infrastructure.[16] In 2011, the value of the smart grid market in China reached $22.3 billion.[17] As of 2012, China had installed 139 million smart meters, enough to cover 35 percent of households.[18]

Other countries in Asia are also investing in smart grid technologies and deployments. South Korea, as of February 2012, had deployed smart meters to fewer than a million households, or roughly 4 percent.[19] But the government plans to install smart meters in half of Korean households by 2016 (10 million units) and in all households by 2020.[20] Japan is already home to one of the most efficient electricity grids in the world, with distribution losses averaging 4.9 percent over the period 2000 to 2010.[21] After the 2011 Fukushima nuclear accident and the subsequent shutdown of most of Japan's nuclear capacity, the government included smart grid infrastructure in its revised energy plan.[22] The country's largest power company, TEPCO, plans to put 27 million smart meters in place between 2014 and 2023, with a median 2018 target of 17 million installations.[23] The complete rollout would cover 38.5 percent of Japanese households.[24] Japan's smart grid market in 2012 was valued at $1 billion.[25]

In the European Union (EU), Electricity Directive 2009/752/EC mandates member states to deploy smart metering systems in 80 percent of households by 2020, where cost-benefit analyses of smart meters are positive.[26] Progress varies from country to country, but as of 2011 an average of 10 percent of EU households had smart meters installed.[27] In addition to directives, the European Commission established the European Electricity Grid Initiative, a nine-year, €2 billion research and development program for smart grid technology and market innovations.[28]

Some European countries have made considerable efforts to develop smart grid networks. Nearly all Italian households have smart meters installed, for example.[29] Italy's advanced metering infrastructure rollout began in 2001 with the commencement of the Telegestore project, the main objective of which was to reduce high non-technical losses on the grid.[30] Smart meter installation was mandated in Italy with Regulatory Order No. 292/06 in 2006.[31] In contrast, East European countries have seen little investment in or development of smart grid networks due to budget constraints.[32]

The United Kingdom plans to begin nationwide installation of smart meters in 2014.[33] The country plans to install these meters in all households by 2019.[34] But in 2012, installations by individual utilities totaled around 540,000, or less than 2 percent of households.[35]

Smart grid investment in Latin America remains generally low. Brazil is an exception, investing $240 million in stimulus funds in 2010.[36] At the end of 2012, Brazil formalized a regulatory framework for smart grid deployment.[37]

Grid-scale energy storage technologies are another important aspect of evolving grid networks, providing an alternate or complementary solution for the integration of variable renewable energy into the grid, among other benefits. In 2010, the value of the global grid-scale energy storage market was $1.5 billion.[38] Installed storage capacity in 2011 totaled 125.5 gigawatts (GW) worldwide.[39] Pumped hydro storage accounted for 98 percent (123.4 GW) of that total.[40] Other means of storing electricity include thermal energy storage, batteries, and compressed air.[41] (See Figure 3.)

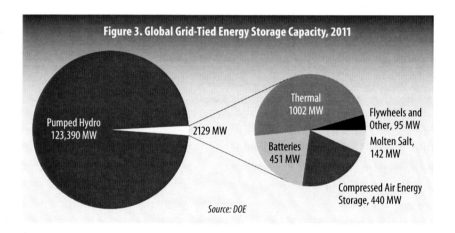

Figure 3. Global Grid-Tied Energy Storage Capacity, 2011

Pumped Hydro 123,390 MW

2129 MW

Thermal 1002 MW

Batteries 451 MW

Flywheels and Other, 95 MW

Molten Salt, 142 MW

Compressed Air Energy Storage, 440 MW

Source: DOE

There were 714 grid-scale energy storage projects worldwide in 2012 in varying stages of operation or development.[42] This represents roughly a 19 percent increase in the number of projects from the previous year.[43] Pumped hydro storage accounted for most capacity additions, with 10,359 megawatts installed from 2007 to 2012.[44] However, new technologies, such as advanced batteries, are expected to play a larger role in the energy storage market in the coming years.[45]

Smart grid networks and energy storage technologies are gaining traction in energy sector development plans, with larger-scale projects beginning or planned for the near future. The next few years will see numerous nationwide smart grid deployment projects and advances in energy storage markets, the success of which will surely influence the respective paths of each technology's development.

Fossil Fuel and Renewable Energy Subsidies on the Rise

Alexander Ochs, Eric Anderson, and Reese Rogers

A recent projection places the total value of conventional global fossil fuel subsidies between $775 billion and more than $1 trillion in 2012, depending on which supports are included in the calculation.[1] In contrast, total subsidies for renewable energy stood at $66 billion in 2010, although that was a 10 percent increase from the previous year.[2] Two thirds of these subsidies went to renewable electricity resources and the remaining third to biofuels.[3]

Although the total subsidies for renewable energy are significantly lower than those for fossil fuels, they are higher per kilowatt-hour if externalities are not included in the calculations. Estimates based on 2009 energy production numbers placed renewable energy subsidies between 1.7¢ and 15¢ per kilowatt-hour (kWh) while subsidies for fossil fuels were estimated at around 0.1–0.7¢ per kWh.[4] Unit subsidy costs for renewables are expected to decrease as technologies become more efficient and the prices of wholesale electricity and transport fuels rise.[5]

Globally negotiated efforts to reduce fossil fuel subsidies have been hindered by competing definitions of subsidies. Calculation methods also vary. The common price gap approach to calculating consumption subsidies uses the difference between the observed domestic prices of energy and the world market prices as an estimate of the impacts of a country's policies on market prices.[6] Some oil exporters, however, argue that production cost rather than market price should be used as the baseline.[7] The difficulties in accurately measuring data are compounded by the lack of transparency among countries with regard to energy subsidies.[8]

The more conventional calculations of fossil fuel subsidies do not take "hidden subsidies" into account. The production and consumption of fossil fuels add costs to society in the form of detrimental impacts on resource availability, the environment, and human health. These costs are not reflected in fossil fuel prices. Monetizing and factoring them into the data would raise fossil fuel subsidies by hundreds of billions of dollars. The U.S. National Academy of Sciences estimates, for example, that fossil fuel subsidies cost the United States $120 billion in pollution and related health care costs every year.[9] And according to one calculation, the United States spends at least $1.6 trillion annually on maintaining the country's fossil-fuel-based transportation infrastructure, including highways and airports.[10] On the global scale, estimates placed the costs of the impacts of greenhouse gases alone at $4.5 trillion in 2008.[11]

In general, traditional calculations account for two kinds of energy subsidies. Production subsidies lower the cost of energy production through preferential tax treatments and direct financial transfers (grants to producers and preferential loans). Consumption subsidies lower the price that consumers pay for energy, usually through tax breaks or underpriced government energy services.

Alexander Ochs heads the Climate and Energy Program at Worldwatch Institute. Eric Anderson was an intern and Reese Rogers is a MAP Sustainable Energy Fellow in this program.

About two thirds of global fossil fuel subsidies are consumption subsidies.[12] In 2010, consumption subsidies in developing countries alone equaled roughly $409 billion, up from $312 billion in 2009 but lower than the $558 billion in 2008.[13] The fluctuation was almost entirely due to variability in fossil fuel prices rather than policy changes. Consumption subsidies in industrial countries averaged $45 billion annually from 2008 to 2010.[14] Figure 1 breaks the global total down into consumption subsidies by source.[15] In developing countries, roughly $193 billion, or 47 percent of all fossil fuel consumption subsidies, went to oil in 2010.[16] Natural gas consumption there received $91 billion in support.[17] Coal received only $3 billion in direct consumption subsidies in these countries, but another $122 billion went to public underpricing of electricity, much of which is generated from burning coal.[18] Consumption subsidies in industrial countries were calculated using a broader definition of support. In 2010, support for oil in industrial countries was valued at roughly $28 billion.[19] Natural gas support in these countries totaled around $10 billion.[20] Coal was supported the least in industrial countries, with $5 billion in subsidies.[21]

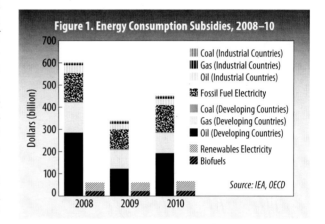

Figure 1. Energy Consumption Subsidies, 2008–10

Coal (Industrial Countries)
Gas (Industrial Countries)
Oil (Industrial Countries)
Fossil Fuel Electricity
Coal (Developing Countries)
Gas (Developing Countries)
Oil (Developing Countries)
Renewables Electricity
Biofuels

Source: IEA, OECD

The rates of subsidization are highest among countries in the Middle East and North Africa that are net exporters of oil and gas.[22] Table 1 shows the 10 highest consumption subsidy rates by country, with the percentage showing the portion of the international market price that is covered by the subsidy. Since 2007, roughly 80 percent of spending on consumption subsidies occurred in countries that are net exporters of fossil fuels.[23]

Production subsidies are more difficult to calculate and are implemented through a wide variety of policies. The most common form is foregone government revenue through such programs as reduced royalty payments and accelerated depreciation.[24] Estimates are that global production subsidies total roughly $100 billion per year.[25] The Organisation for Economic Co-operation and Development (OECD) examined 24 of its 34 member countries and calculated that total production subsidies ranged between $45 billion and $75 billion.[26] Within those countries, subsidies for coal production accounted for roughly 39 percent, making the coal sector the largest beneficiary of fossil fuel production subsidies.[27] Petroleum and natural gas both received roughly 30 percent of fossil fuel production subsidies in OECD countries in 2010.[28]

According to projections by the International Energy Agency (IEA), if fossil fuel subsidies were phased out by 2020, global energy consumption would be 3.9 percent below the figure expected that year if subsidy rates are unchanged.[29] Oil demand would be reduced by 3.7 million barrels per day, natural gas demand would be cut by 330 billion cubic meters, and coal demand would drop by 230 million tons of coal equivalent.[30] And the effects of the subsidy removal would extend beyond the end of the phaseout period. By 2035, oil demand would de-

Table 1. Fossil Fuel Consumption Subsidy Rates as Share of World Market Price, Top 10 Countries, 2010

Country	Share of World Market Price Covered by Subsidy
	(percent)
Kuwait	85.5
Iran	84.6
Saudi Arabia	75.8
Qatar	75.3
Venezuela	75.3
Libya	71.0
UAE	67.8
Turkmenistan	65.1
Algeria	59.8
Uzbekistan	57.1

Source: IEA, World Energy Outlook 2011, at www.iea.org/subsidy/index.html.

crease by 4 percent, natural gas demand by 9.9 percent, and coal demand by 5.3 percent, compared with the baseline projection.[31] Overall, carbon dioxide emissions would be reduced by 4.7 percent in 2020 and 5.8 percent in 2035.[32] The IEA's chief economist recently estimated that eliminating all subsidies in 2012 for coal, gas, and oil could have saved as much as Germany's annual greenhouse gas emissions by 2015, while the emission savings over the next decade might be enough to cover half of the carbon savings needed to stop dangerous levels of climate change.[33]

Progress toward a complete phaseout has been minimal. The 2009 pledge by the Group of 20 (the G20) major economies to reduce "inefficient fossil fuel subsidies" has been left vague and unfulfilled.[34] (And there is no set definition of "inefficient"; countries are left on their own to determine if their subsidies are inefficient.) As of June 2012, G20 countries had not taken any substantial action in response to the pledge: six members opted out of reporting altogether (an increase from two in 2010), and no country has yet initiated a subsidy reform in response to the pledge.[35] Furthermore, there continues to be a large gap between self-reported statistics and independent estimates in some countries.[36]

Nevertheless, some countries have had some success in reducing consumption subsidies. Iran has implemented a radical subsidy reduction program that will bring gasoline prices to no less than 90 percent of the Persian Gulf FOB price and that will distribute half of the money saved to the poorest 80 percent of the population through cash handouts.[37] Nigeria completely eliminated subsidies, but that resulted in widespread protests and violence, so it reinstituted two thirds of the supports.[38]

Some argue that reducing subsidies would disproportionately affect the poor. An IEA survey of 11 developing and emerging countries found that only 2–11 percent of subsidies went to the poorest 20 percent of the population, showing that subsidies tend to be regressive.[39] Subsidies for liquefied petroleum gas, gasoline, and diesel benefit the bottom 20 percent income group the least, while kerosene subsidies generally have the greatest impact.[40] These kerosene subsidies, however, often undermine the competitiveness of renewable alternatives, such as solar lanterns and improved cookstoves, in impoverished communities.[41]

In summary, fossil fuel subsidies continue to far outweigh support for renewable energy. Although independent reporting on these subsidies has increased, global efforts to move forward with subsidy reform have been hindered by a variety of causes, leaving international pledges unfulfilled.

Continued Growth in Renewable Energy Investments

Evan Musolino and Xing Fu-Bertaux

Emerging from the global economic recession, investments in renewable energy technologies continued their steady rise in 2011. Total new investments in renewable power and fuels (excluding large hydropower and solar hot water) jumped 17 percent—reaching $257 billion, up from $220 billion in 2010.[1] (See Figure 1.) In a year marked by falling costs for renewable energy technologies, net investment in renewable power capacity was $40 billion greater than investment in fossil fuel capacity.[2] (Through the first half of 2012, however, total investment fell behind the impressive pace set the previous year, attracting slightly under $108 billion compared with nearly $125 billion in the first half of 2011.)[3]

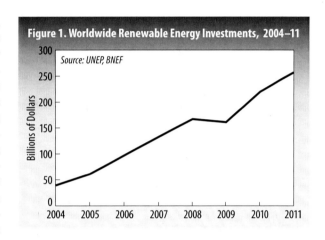

Figure 1. Worldwide Renewable Energy Investments, 2004–11

Source: UNEP, BNEF

Total renewable energy investments in industrial countries in 2011 accounted for 65 percent of global investment, increasing 21 percent to $168 billion overall.[4] In contrast, the 35 percent of global new investment that went to developing countries increased 10 percent, to $89 billion.[5] Of that sum, China, India, and Brazil accounted for $71 billion in total investment.[6] Investment in India grew 62 percent—the highest growth rate for any single country over 2010 totals.[7] In 2011, "financial new investment" in renewable energy installations (a category that excludes small-scale projects and R&D) in industrial countries outpaced investments in the developing world. But in 2010, investments in this category in developing countries had surpassed those in industrial countries for the first time.[8]

A major development in 2011 was the dominance of solar power in technology-specific investments, driven by a 50 percent reduction in price over the year, with $147.4 billion invested in this technology compared with $83.8 billion for wind projects and $10.6 billion for biomass and waste-to-energy technology.[9] (See Table 1.) While this was not the first time solar surpassed wind in total investment, it was the first time that this involved such a wide margin.[10] Biofuels, which held the second overall ranking in technologies as recently as 2006, attracted the fourth highest total investment in 2011 at $6.8 billion, followed by $5.8 billion for small hydro and $2.9 billion for geothermal installations.[11] Marine energy technologies received only $200 million, as they have not been yet commercially deployed.[12]

Investments in small-scale distributed generation power projects (with capaci-

Evan Musolino is a research associate in the Climate and Energy Program at Worldwatch Institute. Xing Fu-Bertaux was a research associate in the program.

Table 1. Renewable Energy Investment by Technology, 2011

Technology	Investment	Change, 2010–11
	(billion dollars)	(percent)
Solar (excluding solar hot water)	147.4	52
Wind	83.8	–12
Biomass and waste-to-energy	10.6	–12
Biofuels	6.8	–20
Small Hydro	5.8	59
Geothermal	2.9	–5
Marine	0.2	–5
Total	257.5	17

Source: U.N. Environment Programme and Bloomberg New Energy Finance, Global Trends in Renewable Energy Investment *(Frankfurt: 2012); REN21,* Global Status Report 2012 *(Paris: 2012).*

Table 2. Renewable Energy Investments, Top Five Countries, 2011

Country	Investment	Change, 2010–11
	(billion dollars)	(percent)
China	52	17
United States	51	57
Germany	31	–12
Italy	29	38
India	12	62

Source: U.N. Environment Programme and Bloomberg New Energy Finance, Global Trends in Renewable Energy Investment *(Frankfurt: 2012); REN21,* Global Status Report 2012 *(Paris: 2012).*

ties of less than 1 megawatt) grew by 25 percent to $75.8 billion in 2011.[13] Italy led all countries in investment in this category at $24.1 billion, outpacing Germany's $20 billion.[14] Japan ($8.1 billion), the United States ($4.2 billion), and Australia and the United Kingdom ($3.8 billion each) round out the top five spots.[15]

Large-scale hydro (with a capacity of more than 50 megawatts) and solar water heaters are not included in the investment statistics presented in Figure 1 and Table 1. Large-scale hydro still constitutes the largest source of renewable electricity installed and generated, although its social and environmental impacts are the subject of debate. Investment in large hydro installations reached $25.5 billion in 2011.[16]

The solar water heater industry has also continued to grow. According to REN21, estimated total investments in solar hot water exceeded $10 billion in 2011, resulting in 49 gigawatts thermal of new capacity additions.[17] When large-scale hydro and solar water heating are included in the overall investment figure, the total renewable energy investment in 2011 jumps to $293 billion.[18]

China attracted $52.2 billion in new investments in 2011, the largest sum of any country.[19] (See Table 2.) This accounted for nearly 60 percent of the total new investments in developing countries and more than 20 percent of the global total.[20] In terms of the pace of growth, however, the United States scored an impressive 57 percent growth in investment over 2010 levels, outpacing all countries except India's 62 percent.[21] Overall, the United States ranks second in total national investment at $50.8 billion, followed by Germany with a total investment of $31 billion.[22]

The International Energy Agency (IEA) projects that 90 percent of the growth in global energy demand during the next 25 years will come from countries that are not members of the Organisation for Economic Co-operation and Development.[23] Investments in renewable energies already constitute the major part of "climate finance" funds designed to help developing countries meet development challenges. Significant new investment in cleaner sources of energy will be required to reduce the share of fossil fuels in the world's total primary energy consumption in order to keep greenhouse gas emissions low enough to maintain the global temperature change within a 2-degrees-Celsius warming scenario.[24] (See Table 3.)

Renewable energy technologies can enhance access to reliable, affordable, and clean modern energy services. They are particularly well suited for remote rural populations, and in many instances they can provide the lowest-cost option for energy access.[25] According to IEA estimates, $48 billion per year is needed to provide universal modern electricity access by 2030.[26]

Renewable energy investments are made by a variety of different actors in a number of value chains. Investments are made to support technology development, equipment manufacturing, and renewable energy projects themselves. Within these value chains, total investments are further divided into a number of different categories.[27] (See Table 4.)

Total R&D investment in renewable energy technologies fell 16 percent to $8.3 billion in 2011.[28] More than half of all publicly funded R&D for renewable energy technologies in 2011 ($2.4 billion) came from special economic stimulus packages. Because many of these programs in countries like Japan and South Korea are expiring, total public R&D for renewables fell 13 percent to $4.6 billion in 2011.[29] Private-sector support for R&D lagged behind public support for the second year in a row, falling 19 percent to $3.7 billion.[30] Overall, R&D investment was reduced for all technologies and nearly all regions, driven by government austerity and caution on the part of the private sector, with only Brazil posting modest gains.[31] Solar technologies constitute the largest share of worldwide R&D, followed by biofuels and wind.[32] R&D trends are generally indicators of the long- to mid-term expectations for the sector, since investments usually start to pay off in several years' time.[33]

Investments in venture capital and private equity fell by 6 percent to $5 billion in 2011.[34] Venture capital, a type of private equity capital typically provided for high-potential technology companies in the early market deployment phase, rose 5 percent to $2.5 billion, but private equity investment fell for the third consecutive

Table 3. Average Annual Investments in Renewable Energy Needed by 2020 to Keep Global Temperature Rise to 2 Degrees Celsius

Technology	Investment
	(billion dollars)
Hydro	80
Onshore wind	60
Solar photovoltaics	50
Concentrating solar power	15
Offshore wind	10
Bioenergy	10
Geothermal	10
Total	235

Source: International Energy Agency, Tracking Clean Energy Progress (Paris: 2012).

Table 4. Worldwide New Renewable Energy Investments, Excluding Large-Scale Hydro, 2011

Category of Investment	Investment
	(billion dollars)
Venture capital and asset investments	160.8
Small-scale distributed	75.8
Equipment manufacturing (including private equity)	12.6
R&D (corporate and government)	8.3
Total	257.5

Source: U.N. Environment Programme and Bloomberg New Energy Finance, Global Trends in Renewable Energy Investment (Frankfurt: 2012).

year to $2.5 billion, a decrease of 15 percent.[35] Investments in venture capital are usually a good indicator of near-term expectations for the sector.

Asset financing—an indicator of current sector activity—amounted to $164 billion in 2011 (including reinvested equity, which does not qualify as new investment and is not included in Table 4), an increase from $139 billion in 2010.[36] It accounted for 64 percent of total new renewable energy investment.[37] China's $49.7 billion, a 20 percent increase from 2010, accounted for 30 percent of total asset financing in 2011.[38] An impressive 96 percent yearly growth rate helped the United States outpace Europe for second place at $40.9 billion.[39] Europe now holds third place with $40.8 billion.[40] In terms of technologies, the wind sector was again the leader in asset financing—attracting $82.4 billion, slightly over half of the total.[41] Solar asset financing totaled $62.1 billion, a one-year growth of 147 percent, which helped to close the significant gap between technologies during 2011.[42] Small hydropower also posted strong asset finance growth, increasing from $2.8 billion to $5.4 billion over the year, while biofuels, biomass, and geothermal were all victims of diminished asset finance investment in 2011.[43]

Auto Production Roars to New Records

Michael Renner

Following a plunge in output triggered by the global economic crisis, world auto production came roaring back to new peaks. According to London-based IHS Automotive, passenger-car production rose from 60.1 million in 2010 to 62.6 million in 2011—and 2012 may have brought a new all-time record of 66.1 million.[1] (See Figure 1.) Even though output of light trucks has declined, the combined numbers for passenger vehicles rose from 74.4 million in 2010 to 76.8 million in 2011 and may have surpassed 80 million in 2012.[2]

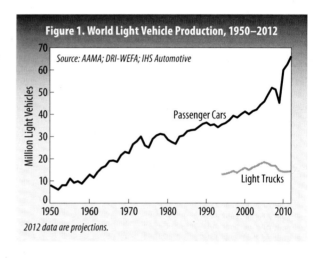

Figure 1. World Light Vehicle Production, 1950–2012

Source: AAMA; DRI-WEFA; IHS Automotive

2012 data are projections.

The auto industry's production capacities are far from fully used. Pricewaterhouse-Coopers (PWC) estimates current global vehicle assembly capacity at almost 95 million.[3] Auto manufacturing capacity continues to grow, and annual output could reach the 100 million mark by 2016.[4]

Global sales of passenger vehicles increased from 75.4 million in 2010 to 78.6 million in 2011, with a projected 81.8 million in 2012.[5] Rising sales numbers translate into ever-expanding fleets. An estimated 691 million passenger cars were on the world's roads in 2011.[6] When both light- and heavy-duty trucks are included, the number rises to 979 million vehicles, which was 30 million more than just a year earlier.[7] By the end of 2012, the number may have topped 1 billion vehicles—one for every seven people on the planet.[8] One of the main drivers behind this growth is China, where the passenger vehicle fleet grew at an annual average rate of 25 percent in 2000–11, from under 10 million cars to 73 million.[9]

The top four producers of light vehicles—China, the United States, Japan, and Germany—together account for more than half of global output.[10] (See Figure 2.) China consolidated its lead, manufacturing 17.3 million light vehicles in 2011, with 2012 output projected at 18.4 million.[11] U.S.-based manufacturers continue to recover from the massive crisis of 2008–09, producing 8.5 million vehicles in 2011, a number that could have risen by another million in 2012.[12] Japan's auto industry is slowly recovering from the March 2011 earthquake and tsunami disaster that forced plant shutdowns.[13] Its output remains more than 3 million units below the peak level of 2007 and is just below that of the United States.[14] Germany's production has held far steadier than that of the other top manufacturers.

Michael Renner is a senior researcher at Worldwatch Institute.

Figure 2. Light Vehicle Production, Leading Countries, 1995–2012

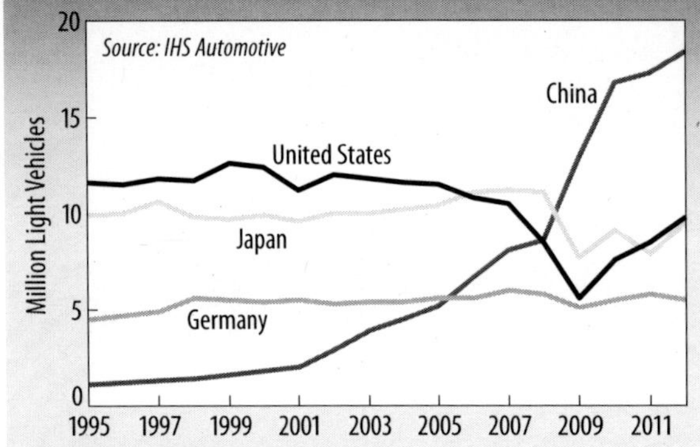

Source: IHS Automotive

2012 data are projections.

Table 1. Top 10 Light Vehicle Manufacturers, 2011

Alliance Group	Home Country	Production
		(million vehicles)
General Motors	United States	9.2
Volkswagen	Germany	8.5
Toyota	Japan	8.5
Renault-Nissan	France	8.0
Hyundai	South Korea	6.7
Ford	United States	5.4
Fiat	Italy	4.2
PSA	France	3.6
Honda	Japan	3.0
Suzuki	Japan	2.4

Source: PWC, "Autofacts: Quarterly Forecast Update," January 2012.

Driven strongly by export sales, South Korea's production continues to rise, closing in on that of Germany and reaching 4.6 million units in 2011.[15] It is followed by India (3.6 million) and Brazil (3.2 million).[16] Mexico, Spain, France, and Canada are all manufacturing between 2.1 million and 2.6 million cars each, and Russia rounds off the top dozen with 1.9 million units.[17] Another five countries—Iran, Thailand, the United Kingdom, Turkey, and the Czech Republic—each produce more than 1 million light vehicles annually.[18]

The top 10 manufacturing companies are headquartered in just six countries but have factories around the world. They account for 80 percent of global light vehicle production.[19] The GM Group led in 2011, with 9.2 million vehicles assembled, followed by VW and Toyota with 8.5 million each.[20] (See Table 1.) The top 20 companies control 94 percent of global assembly.[21] This larger group includes two more companies each from Germany and Japan, India's Tata Group, the Chinese firms Chang'an, Chery, Geely, and Great Wall, and Iran's SAIPA.[22]

Automobiles are major contributors to air pollution and greenhouse gas emissions. Greater fuel efficiency (along with the use of cleaner fuels) can help mitigate these impacts, although increases in the numbers of cars and the distances driven threaten to overwhelm fuel economy advances.

Fuel efficiency has been improving in all the major car nations over the past decade, and stricter consumption limits for coming years have been enacted or proposed.[23] (See Figure 3.) Japan and the European Union (EU) continue to be the global leaders; South Korea has improved its fuel economy by one third since 2003; and China is considering a limit of 5 liters per 100 kilometer (km) for 2020 that would bring it close to Japan's 4.5 liters per 100 km standard for 2020.[24] The United States, Canada, and Australia are also making progress, but nonetheless continue to lag behind.[25] For example, the Obama administration's limits for 2025 represent the most ambitious step ever taken in the United States on fuel efficiency, but they are similar to what Japan already requires for 2015.[26]

In order to force fuel efficiency improvements, the EU has enacted binding limits on how much carbon dioxide (CO_2) a car may emit per kilometer travelled. By 2015, car manufacturers must meet a fleet average of 130 grams per kilometer (g/km).[27] Still, this represents a much delayed and watered-down goal compared with a limit of 120 grams by 2005, as originally proposed by European environment ministers nearly 20 years ago, in 1994, and it is expected that loopholes will effectively allow an average of 140 g/km.[28] Because 140 grams was the average already achieved by new cars in 2010 (down from 175 grams in 2000), the current EU limits are too soft to be an effective driver of further improvements.[29]

In the EU market, the best performers in 2010 were Fiat (126 g/km), Toyota (130), and PSA (131). Daimler (161) and Volvo (157) were at the other end of the spectrum.[30] Among EU member states, new car registrations in 2010 in Denmark and Portugal had the lowest CO_2 footprint (with an average of 127 g/km), followed by France (131) and Italy (133).[31] Germany's vehicle footprint, at 151 g/km, was much higher, although topped by Estonia and Latvia (162 each).[32]

Light vehicles purchased in the United States averaged emissions of 243 g/km in 2011, down 43 percent from 423 grams in 1975.[33] In the U.S. market, the Korean firms Hyundai and Kia performed best in model year 2011 (201 and 203 g/km, respectively), followed by the Japanese companies Honda, Toyota, and Mazda, as well as Germany's VW.[34] At 280 g/km, Chrysler is the worst performer in the U.S. market, along with the German company Daimler (278 g/km).[35]

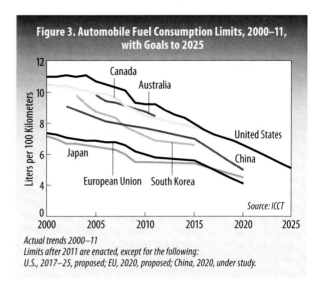

Figure 3. Automobile Fuel Consumption Limits, 2000–11, with Goals to 2025

Source: ICCT

Actual trends 2000–11
Limits after 2011 are enacted, except for the following:
U.S., 2017–25, proposed; EU, 2020, proposed; China, 2020, under study.

Table 2. Types of Vehicles Produced, 2011

Type of Vehicle	Production	
	(million vehicles)	(percent)
Combustion, Gasoline	57.74	78.0
Combustion, Diesel	14.96	20.2
Gasoline-Electric Hybrid	1.22	1.6
Electric	0.15	0.2
Total	74.07	100

Source: PWC, "Autofacts: Quarterly Forecast Update," January 2012.
Note: PWC's 2011 production figure is lower than IHS Automotive's estimate of 76.8 million.

The vast majority of light vehicles produced have conventional types of propulsion systems, either gasoline or diesel-powered combustion engines.[36] (See Table 2.) Hybrid vehicles are growing in number, but they remain below 2 percent of total vehicle output.[37] (Hybrid-electric vehicles combine an internal combustion engine and an electric motor, along with a generator and battery). In 2011, just over 400,000 Toyota Priuses, by far the best-selling hybrid, were purchased (253,000 in Japan; 137,000 in the United States; and 26,000 in Europe).[38] Altogether Toyota has sold 4 million hybrids since 1997, of which the Prius accounts for 2.9 million.[39]

Electric vehicle (EV) production is still at barely perceptible levels. A number of countries have issued targets for future EV fleets, but it remains to be seen whether these goals can be met. For instance, China wants to put 5 million plug-in hybrid electric and fully electric vehicles on its roads by 2020—which could account for more than 40 percent of a future global EV fleet that year.[40] (Plug-in hybrid electric vehicles can be powered either with electricity or fossil fuels. Electric vehicles do not rely on any internal combustion system.) The government envisions that annual sales will rise from 8,159 in 2011 to 1.58 million in 2020.[41] An analysis by Deutsche Bank Climate Advisors suggests that production of 1.1 million and a fleet of 3.5 million EVs in China is a more realistic projection.[42]

While discussions about reducing the environmental impacts of cars tend to focus on technical improvements (engines, aerodynamic design, fuels, etc.), another aspect concerns the distances traveled by car. The combination of large numbers of automobiles on the road and long distances per vehicle has long made the United States the country with by far the highest number of passenger-kilometers (pkm) for private cars. The U.S. total grew from 2.8 trillion pkm in 1970 to 3.7 trillion in 1990.[43] Travel distances grew to a peak level of 4.3 trillion pkm in 2005, but the total has since declined slightly to 4.1 trillion.[44] Even though the United States has just 25 percent of the total population of the Organisation for Economic Co-operation and Development (OECD), in 2008 this one country accounted for slightly above 40 percent of the 10.3 trillion pkm driven in all OECD member countries.[45]

In Japan, Germany, the United Kingdom, and Italy, car travel distances all accelerated in the second half of the 1980s and the early 1990s, although in the case of Germany this principally reflected reunification with the former East Germany.[46] (See Figure 4.) With the exception of Germany, the pkm figure leveled off after 2000 in these countries and in France.[47] Japan saw a 6 percent decline in pkm between 1999 and 2008, and South Korea registered a 35 percent decline between 2001 and 2008.[48] In sharp contrast, the distance traveled by Polish drivers soared 6.4-fold between 1985 and 2010.[49]

The Chinese are also increasingly taking to the roads, with driving distances rising from 262 billion pkm in 1990 to 1.4 trillion pkm in 2009, slightly more than a fivefold expansion.[50] Car travel in non-OECD countries doubled between 1975 and 2000, but then it picked up pace by doubling again in just the decade to 2010.[51] The International Transport Forum notes that global light-duty vehicle use was nearly 2.5 times higher in 2010 than it was in 1975.[52]

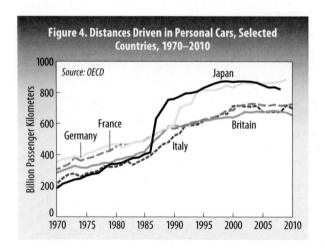

Figure 4. Distances Driven in Personal Cars, Selected Countries, 1970–2010

On a per capita basis, people in OECD countries drive about 8,500 kilometers in private cars annually.[53] People in Canada (14,600 km) and the United States (13,500 km) drive greater distances than people in Europe (an average of just over 11,000 km in the four largest European countries).[54] In Japan, the average driver covers only about 6,400 km, a testament to that country's excellent and popular intercity rail network.[55] And in China, the average distance per person works out to a much shorter 1,000 km.[56]

Environment and Climate Trends

Opening of CO_2-to-methanol plant at Svartsengi Geothermal Power Plant, Reykjanes, Iceland

ThinkGeoEnergy

For additional environment and climate trends, go to vitalsigns.worldwatch.org.

Carbon Dioxide Emissions and Concentrations on the Rise as Kyoto Era Fades

Xing Fu-Bertaux

According to on-site measurements by the Scripps Institute of Oceanography, global atmospheric carbon dioxide (CO_2) concentrations reached 391.3 parts per million (ppm) in 2011, up from 388.56 ppm in 2010 and from 280 ppm in pre-industrial times.[1] (See Figure 1.) Carbon dioxide accounts for more than 70 percent of the greenhouse gases (GHGs) in the atmosphere and—thanks to its very long life span—is the most important anthropogenic greenhouse gas responsible for global warming.[2]

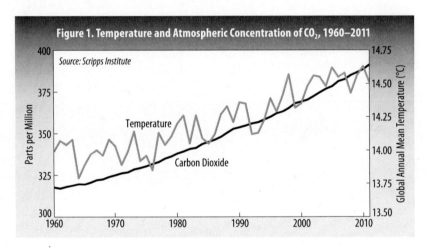

Figure 1. Temperature and Atmospheric Concentration of CO_2, 1960–2011

After declining 1.5 percent in 2009, global CO_2 emissions jumped 5.8 percent in 2010, an unprecedented increase in the last two decades.[3] CO_2 levels are now 45 percent above the 1990 level, the reference base year under the United Nations Framework Convention on Climate Change (UNFCCC).[4] Levels of methane (CH_4) and nitrous oxide (N_2O) have also increased significantly, but they account for a smaller share of greenhouse gases in the atmosphere—17 percent and 8.7 percent, respectively.[5]

Deforestation and logging, forest and peat fires, and the decomposition of organic carbon drained in peat soils are estimated at around 14 percent of global CO_2 emissions; however, this number is highly uncertain and varies from 15 percent to 30 percent between years.[6] Industrial processes, mainly the production of cement, constitute another 5 percent of global CO_2 emissions.[7]

The energy sector represents the largest source of CO_2 emissions worldwide.

Xing Fu-Bertaux was a research associate in the Worldwatch Institute's Climate and Energy program.

(See Figure 2.) More than 70 percent of these emissions result from burning fossil fuels for electricity generation, transportation, manufacturing, and construction.[8] In 2009, some 41 percent of energy-related CO_2 emissions came from electricity generation and heating.[9] (See Figure 3.) Another 23 percent were produced by road, air, and marine transportation; 20 percent came from energy used in the industrial sector; and the residential sector accounted for 6 percent of energy-related emissions.[10]

With the economic recovery, oil consumption increased by 3.3 percent in 2010, natural gas by 7.4 percent, and coal consumption by 7.4 percent.[11] Emissions from the combustion of these fossil fuels rose too. In 2010, coal combustion constituted 40 percent of energy-related CO_2 emissions, while oil represented 37 percent and natural gas 20 percent.[12] Burning coal generates about twice as much CO_2 as gas and oil do because of the larger carbon content per unit of energy released.[13]

The year 2010 was marked by a general growth in CO_2 emissions in developing countries as well as richer industrial ones.[14] (See Table 1.) Emissions in countries that do not belong to the Organisation for Economic Co-operation and Development (OECD) grew by 7.6 percent that year, pulled by a significant increase in almost all the parts of the developing

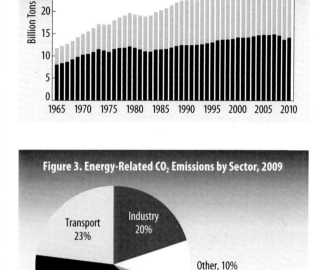

Figure 2. CO_2 Emissions from Energy Use, 1965–2010

Figure 3. Energy-Related CO_2 Emissions by Sector, 2009

world.[15] After a slump of 6.5 percent in 2009, emissions of OECD countries soared again and grew by 3.4 percent.[16] Overall, however, Annex I countries—which includes most OECD countries that were assigned internationally legally binding emissions reduction targets and all economies in transition—reduced their emissions from 1990 levels, mainly due to the deep cuts in emissions in the transition economies, including Russia, Ukraine, and other East European countries.[17]

Annex I countries were expected to meet the 4.6 percent reduction mandated by the Kyoto Protocol by the end of 2012.[18] But there are large national differences among them: some countries, such as Australia, Canada, New Zealand, and Spain, will not meet their reduction targets without buying additional credits from other countries.[19] The United States, which signed but never ratified the Kyoto Protocol, will also be unable to meet its original reduction target of 6 percent, as its greenhouse gases have increased by 12.9 percent since 1990.[20] Since the 17th Conference of the Parties to the UNFCCC in December 2011, Japan, Russia, and Canada have decided not to take on additional emissions reduction targets in the coming decade.[21] The Kyoto Protocol has become a symbolic international instrument

	2009	2010	Change, 2009–10	Change, 1990–2010	Share of Total
Table 1. Regional Variation in CO_2 from Energy Use, 2009–10 , and Change, 1990–2010					
	(million tons CO_2)			(percent)	
China	7,626	8,415	10.3	236.6	25
Rest of non-OECD Asia	4,164	4,445	6.8	163.7	13
Middle East	1,819	1,913	5.2	156.7	6
Africa	1,045	1,077	3.0	61.5	3
Total non-OECD	17,667	19,018	7.6	86.5	57
United States	5,904	6,145	4.1	12.9	19
European Union (current 27 member countries)	4,055	4,143	2.2	–7.6	12
Japan	1,225	1,308	6.8	13.0	4
Canada	590	605	2.6	22.3	2
Total OECD	13,671	14,141	3.4	13.9	43
World	31,339	33,158	5.8	46.6	100

Source: BP, Statistical Review of World Energy *(London: June 2011).*

Table 2. Top 10 Emitters from Energy Use in 2010

Country	CO_2 Emissions from Fuel Combustion	Share of World Total
	(million tons)	(percent)
China	8,415	25.4
United States	6,145	18.5
India	1,707	5.1
Russia	1,700	5.1
Japan	1,308	3.9
Germany	828	2.5
South Korea	716	2.2
Canada	605	1.8
Saudi Arabia	562	1.7
Iran	557	1.7
Total top 10	22,543	68
World	33,158	100

Source: BP, Statistical Review of World Energy *(London: June 2011).*

that only regulates around 15 percent of global GHG emissions, as now most of the world's major emitters do not have internationally legally binding reduction targets.[22]

In 2010 China was the world's largest emitter of CO_2, followed by the United States, India, and Russia.[23] (See Table 2.) Since 2006, non-OECD countries as a group emit more GHGs each year than OECD countries.[24] But the picture changes when population size is taken into account. (See Table 3.) China, the world's largest emitter, only ranks sixty-first per person, whereas the United States ranks second in absolute terms and tenth per person.[25] For some developing countries, the gap is even wider: while India ranks third in absolute terms, its emissions per person are far below the world average, ranking one hundred and twenty-first per person.[26]

The carbon intensity of gross domestic product (GDP) is a good indicator of the reliance of a country's economy on fossil fuels. Worldwide,

Table 3. Total Energy-Related CO_2 Emission in 2009 and Emissions in Relation to Population and Economy

Country or Region	CO_2 Emissions from Fuel Combustion	Carbon Emissions per Person	Carbon Emissions per GDP PPP
	(million tons)	(tons)	(tons of CO_2 per million dollars PPP)
China	8,415	6.3	550
United States	6,145	20.0	460
European Union	4,143	8.3	290
India	1,707	1.5	350
Russia	1,700	12.0	1,000
Japan	1,308	10.3	320
Africa	1,077	1.1	360

Source: BP, Statistical Review of World Energy (London: June 2011); population and GDP data from IEA, CO_2 Emissions from Fuel Combustion, Highlights (Paris: 2011), pp. 82, 85.

the highest levels of emissions per unit of GDP are observed in the Middle East, where economic revenue is highly dependent on oil production and exports, and in countries of the former Soviet Union, which have energy-intensive industries.[27] But carbon emissions are not necessarily related to a nation's development. For example, Japan's economy is significantly less reliant on CO_2 emissions than the United States is, despite a similar living standard.[28] And emissions per unit of GDP in Russia are three times those in Japan, despite a much lower living standard.[29]

In early 2007, the Intergovernmental Panel on Climate Change (IPCC) released its strongest statement yet linking CO_2 emissions and increasing global temperatures, stating with more than 90 percent certainty that the warming over the past 50 years has been caused by human activities.[30] Growth of CO_2 levels in the atmosphere has been accompanied by significant temperature increase in the past decade: the global average surface temperature in 2011 was the ninth warmest since 1880.[31] NASA scientists established 2000–09 as the warmest decade on record since 1880.[32]

Since 2007, new peer-reviewed science and technology articles suggest that the impacts of climate change in many areas of the world are not advancing linearly: profound changes are already occurring, and models project even greater changes for the remainder of the twenty-first century.[33] The findings support the need for rapid and deep cuts in GHG emissions.[34] The IPCC's recent *Special Report on Managing the Risks of Extreme Events and Disasters to Advance Climate Change Adaptation* warns that a warmer world will likely lead to disruptive changes in the frequency and magnitude of extreme events, such as wildfires, heat waves and cyclones, reinforcing the belief that heat waves, intense rainfall events, category 4 and 5 storms, as well as drying trends will increase across much of the northern and southern hemispheres.[35]

Other articles have taken on an even more alarmist tone. A recent report by the London-based Royal Society estimated that at 4° Celsius of global temperature increase, which is in the middle of the range of current projections, half the world's current agricultural land would become unusable, sea levels would rise by up to two meters, and around 40 percent of the world's species would become extinct.[36]

Carbon Capture and Storage Experiences Limited Growth in 2011

Matt Lucky

Funding for large-scale carbon capture and storage (CCS) projects remained relatively unchanged in 2011, with total funding from governments reaching $23.5 billion.[1] (See Figure 1.) Overall, the number of active and planned large-scale CCS projects declined in 2011, although the total operating storage capacity increased.

In March 2012, the Global CCS Institute identified 75 large-scale fully integrated CCS projects in 17 countries at various stages of development—4 projects fewer than at the end of 2010.[2] Only 8 of these plants are operational, the same number as in 2009 and 2010.[3] (See Figure 2.) These 8 projects store a combined total of 23.18 million tons of carbon dioxide (CO_2) a year (Mtpa), about as much as emitted annually by 4.5 million passenger vehicles in the United States.[4] Operating storage capacity has more than doubled since late 2010.[5]

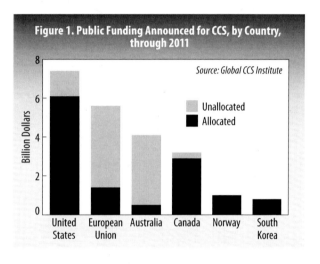

Figure 1. Public Funding Announced for CCS, by Country, through 2011

Source: Global CCS Institute

At the end of 2011, the United States remained the largest funder of large-scale CCS projects ($7.4 billion), having allocated $6.1 billion to projects and with an additional $1.3 billion set aside for future projects.[6] The European Union has announced the next largest amount of funding ($5.6 billion), although Canada has actually allocated more money to date ($2.9 billion).[7] In March 2011, the United Kingdom decided to no longer pursue a CCS Electricity Levy; instead, general taxes will be used to fund that nation's CCS projects.[8]

Although the number of operating plants has not changed since 2009, an additional facility—the Century Plant in the United States—began operation in 2010 and added 3.5 Mtpa more storage capacity in 2012 to reach a total storage capacity of 8.5 Mtpa.[9] At the same time, the Rangeley Weber and Salt Creek enhanced oil recovery plants were reclassified as one plant, as they share a single CO_2 capture source, which kept the number of operating plants at 8.[10]

Construction began on 4 large-scale CCS projects since the end of 2010.[11] There are now 7 large-scale CCS plants currently under construction, bringing the total annual storage capacity of operating and under-construction plants to 34.97 Mtpa.[12] If the remaining 60 projects under planning or development are built, they would add an additional 134.25 Mtpa of capacity.[13] The total storage capacity of all active and planned large-scale CCS projects is 169.2 Mtpa, equivalent to only

Matt Lucky is a Climate and Energy research associate at Worldwatch Institute.

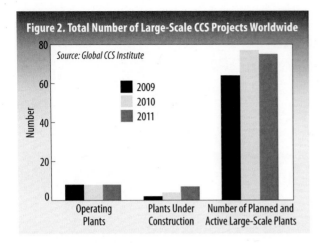

Figure 2. Total Number of Large-Scale CCS Projects Worldwide

Source: Global CCS Institute

about 0.5 percent of global emissions from energy use in 2010.[14]

A total of 13 projects were canceled or postponed in 2011.[15] In most cases these were ruled uneconomical.[16] Other reasons were also cited, however: the Jänschwalde project in Germany was canceled due to local community opposition to the chosen storage site.[17]

Governments and industry have continued to invest heavily in CCS with the aim of substantially decreasing CO_2 emissions and combating climate change. Funding for CCS is mostly targeted at fossil fuel power plants, especially greenhouse gas–intensive coal plants, although CCS can also be used in natural gas power plants and many industrial facilities. It is estimated that CCS can cut CO_2 emissions from coal-fired power plants by 85–95 percent from what they would otherwise be.[18]

In March 2012, the U.S. Environmental Protection Agency implemented regulations on CO_2 emissions from power plants.[19] As a result, U.S. power producers will be unable to build traditional coal plants without carbon controls—including CCS—in the future.[20] For the United States and many other countries that want to curb greenhouse gas emissions while also continuing to burn coal, CCS will likely become an increasingly important technology.

According to the International Energy Agency (IEA), an additional $2.5–3 trillion will need to be invested in CCS between 2010 and 2050 to cut greenhouse gas emissions in half by mid-century; this scenario envisions the completion of 3,000 large-scale CCS plants by then.[21] On average, $5–6.5 billion a year will need to be invested in CCS globally until 2020 for the development of this technology.[22] Although a majority of planned and active CCS projects are in North America and Europe (see Figure 3), the IEA believes greater investment in CCS projects will be needed in the developing world in the future.[23]

About 76 percent of government funding for large-scale CCS has been allocated to power generation projects.[24] Although 43 large-scale power plant CCS projects are currently under development—2 of which are being built in Canada and the United States—no large-scale commercial power plants with CCS technology were operating as of March 2012.[25] Of the 8 operating plants with CCS technology, 6 are for natural gas processing, 1 is for synthetic natural gas production, and 1 is for fertilizer production.[26]

CCS has three steps: capturing CO_2 from a source such as a power plant's flue gas, moving this CO_2 to a storage site, and injecting it into a storage reservoir.[27] Today, there are three primary methods for that first step in power plants: post-combustion, pre-combustion, and oxy-fuel combustion.[28] As of now, none of these technologies has emerged as the clear winner in terms of cost or feasibility.

For power generation, pre-combustion and post-combustion technologies have attracted similar levels of investment: $3.5 billion and $3.3 billion, respectively.[29]

Investments in oxy-fuel CCS are significantly smaller at $1.7 billion.[30] When considering all industries, including power generation, natural gas processing, oil refining, fertilizer, and chemical production, pre-combustion technology receives the most funding at about $5.5 billion.[31]

Pre-combustion combines gasification of a solid fuel with CO_2 separation to yield a hydrogen gas, which can then be burned without emitting greenhouse gases. All 8 currently operating large-scale CCS projects use pre-combustion technology, although none of these plants are used primarily for power generation.[32] Five additional plants that are currently under construction use this capture technology, including 1 power plant—the Kemper County IGCC Project in Mississippi.[33] Another 32 pre-combustion projects are currently under development.[34]

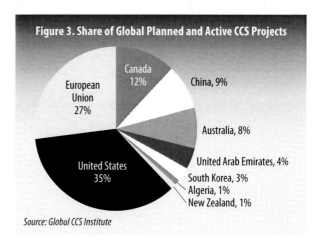

Figure 3. Share of Global Planned and Active CCS Projects

Canada 12%
European Union 27%
China, 9%
Australia, 8%
United Arab Emirates, 4%
United States 35%
South Korea, 3%
Algeria, 1%
New Zealand, 1%

Source: Global CCS Institute

In post-combustion, CO_2 is extracted from flue gases that emerge from the combustion process. Since it requires few changes from combustion technology, it has received comparable funding to pre-combustion from the power industry. There are currently no large-scale post-combustion CCS projects in operation.[35] One plant—the Boundary Dam Integrated Carbon Capture and Sequestration Demonstration Project in Canada—is currently under construction, while an additional 17 are in development stages.[36]

Oxy-fuel technology burns fuel in oxygen mixed with recycled flue gas rather than nitrogen-rich air, producing a CO_2-rich gas that is ready to be stored. There are no operating large-scale oxy-fuel CCS projects. There are currently 5 projects in the development stages in China, the Netherlands, Spain, the United Kingdom, and the United States, all of which are for power generation.[37] The first large-scale oxy-fuel CCS power plant is scheduled to come online in 2015.[38]

Although CO_2 has been transported safely over the past few decades, the scale of future transport infrastructure will require tremendous funding if CCS is to be adopted as a major greenhouse gas mitigation tool.[39] In general, CO_2 is moved to storage sites in pipelines—95 percent of large-scale active and planned CCS projects use or will use pipelines—but vehicles and ships may also be used where pipeline transport is not feasible.[40]

The main options for CO_2 storage are deep saline aquifers and depleted oil and gas fields. Onshore depleted oil and gas fields are the cheapest option (see Figure 4), but a majority of the world's identified storage capacity is found in deep saline aquifers.[41]

To date, however, oil reservoirs have received the greatest investment for carbon storage. Injecting captured CO_2 into oil wells lets producers pull up economically inaccessible oil and add significant time to the life of an oil field, in a process known as enhanced oil recovery. Five of the 8 existing projects inject CO_2 into depleted oil reservoirs for this purpose.[42] This suggests that CCS is being funded not only

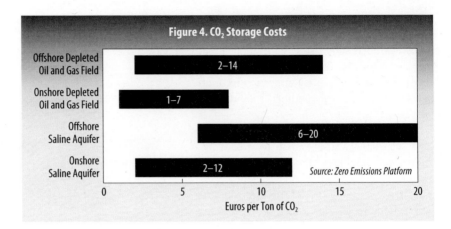

Figure 4. CO₂ Storage Costs

to reduce greenhouse gas emissions but also to extract more-traditional fossil fuels.

The other 3 currently operating plants store CO_2 in saline formations: 2 offshore and 1 onshore.[43] The United States also recently commenced construction of its first large-scale CCS plant that will store CO_2 in a saline formation, the Illinois Industrial Carbon Capture and Sequestration project.[44]

Despite the continued investment in CCS technology, the incremental cost of including CCS in the power sector remains quite high. For countries that belong to the Organisation for Economic Co-operation and Development, a recent IEA study found that adding CCS increased the levelized cost of electricity for coal plants by between 39 and 64 percent, resulting in prices between 10.2¢ and 10.7¢ per kilowatt-hour (kWh).[45] For natural gas plants, electricity costs increased by 33 percent, leading to a price of 10.2¢ per kWh.[46]

Because CCS is considered by many observers as a tool to mitigate greenhouse gas emissions and climate change, it is important to address environmental concerns about its use. Some scientists and industry representatives maintain that CO_2 can be stored safely for hundreds of thousands of years.[47] Recent reports, however, of pond water carbonation and blowouts on some residents' land have given CCS opponents more reason to argue that CO_2 may not in fact be leak-free.[48] Other researchers, though, suggest the CO_2 that was documented came from local wetlands, not the nearby CCS project.[49] Another environmental concern relates to water: CCS significantly increases water usage and can lead to drinking water contamination.[50] A U.S. Department of Energy study suggests that coal plants with carbon capture technology will consume between 87 and 93 percent more water per megawatt-hour than a similar plant without that technology.[51] Carbon capture technology also consumes power, causing relative plant efficiency losses of 15 percent for natural gas power plants and 20–25 percent for coal-fired power plants.[52]

An international regulatory framework for CCS continues to develop slowly. The Convention on the Prevention of Marine Pollution by Dumping of Wastes and Other Matter (known as the London Protocol) was amended twice, for example, to allow offshore CO_2 storage and cross-border transportation.[53] And after many years of stalled negotiations on CCS, international climate change negotiators agreed at the 16th Conference of the Parties (COP) in Cancun, Mexico, that this technology is eligible for use under the Clean Development Mechanism (CDM).[54] This decision, however, proved controversial, as many people viewed it as going against the spirit of the CDM, which is aimed at stimulating sustainable development and emission reductions, as it could potentially prolong the life of carbon-intensive industries.[55] At the 17th COP in Durban, South Africa, countries adopted procedures and modalities concerning CCS in the CDM.[56] Methods for project approval and development as well as project-specific issues, including transboundary concerns, were expected to be discussed at the 18th COP in Qatar in December 2012.[57]

Food and Agriculture Trends

Above the entrance to the Hereford Cattle Society, Herefordshire, United Kingdom

For additional food and agriculture trends, go to vitalsigns.worldwatch.org.

Global Grain Production at Record High Despite Extreme Climatic Events

Danielle Nierenberg and Katie Spoden

In 2012, global grain production was expected to reach a record high of 2.37 billion tons, an increase of 1 percent from 2011 levels.[1] (See Figure 1.) Grain crops are used for human consumption, animal feed, and biofuels. According to the U.N. Food and Agriculture Organization (FAO), the production of grain for animal feed is growing the fastest—a 2.1 percent increase from 2011.[2] Grain for direct consumption by people grew 1.1 percent from 2011.[3] Grain used for biofuel production and other non-feed uses has slowed to a 1 percent increase from 2011 (compared with an 8.2 percent increase from 2008 to 2009).[4]

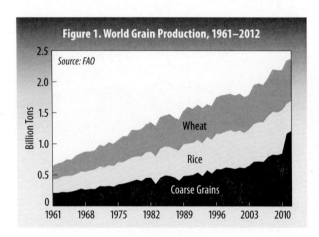

Figure 1. World Grain Production, 1961–2012

Source: FAO

Billion Tons

Wheat

Rice

Coarse Grains

In 2011, the amount of grain used for food totaled 570.7 million tons, with India consuming 89 million tons, China 87 million tons, and the United States 28 million tons, according to the International Grains Council.[5] There is a huge global reliance on wheat, maize (corn), and rice for daily sustenance. Of the 50,000 edible plants in the world, these three grains account for two thirds of the world's food energy intake.[6] Grains provide the majority of calories in diets worldwide. Available caloric intake from grain ranges from 23 percent in the United States to 60 percent in developing Asia and 62 percent in North Africa.[7]

FAO expected global maize production to increase 4.1 percent from 2011, reaching an estimated global production of 916 million tons in 2012.[8] For 2012, rice production was forecast by the FAO at 488 million tons (milled), an increase of 7.9 million tons from 2011.[9] Wheat production was estimated to reach 675.1 million tons in 2012, dropping 3.6 percent from 2011.[10] The decline in wheat production is partially attributed to poor weather during the growing season, including droughts in Morocco and Central Asia and harsh winters in Europe (in Poland, France, Germany, the Czech Republic, Bulgaria, and Hungary).[11]

Maize production in the United States—the largest producer—was expected to reach a record 345 million tons in 2012; however, drought in the Great Plains severely altered this estimate.[12] Maize yields for the 2012–13 growing season are now expected to decrease 13 percent from 2011 production, for a total production of only 274.3 million tons (10.8 billion bushels).[13] Argentina experienced an 11 percent decline in maize production in 2011, producing just 20.3 million tons, also due

Danielle Nierenberg formerly directed the Nourishing the Planet Program at Worldwatch Institute. **Katie Spoden** was a food and agriculture intern at Worldwatch.

to extended drought.[14] Brazil, on the other hand, produced a record-high 66 million tons of maize in 2012, a 17 percent increase from the previous record in 2011.[15]

Global rice production achieved an all-time high in 2011, a 2.6 percent increase to 480.1 million tons (milled rice equivalent) from 2010.[16] With the 2012 rice season just beginning and with farmers south of the equator beginning to harvest, production was forecast to increase 1.7 percent to 488 million tons in 2012.[17] In Thailand, Laos, Myanmar, and Western Africa, rice production was expected to recover after floods in 2011.[18] In Argentina, Brazil, Paraguay, and Uruguay, however, rice cultivation dropped due to lower prices, increased costs, and water shortages.[19] Overall, regional rice production in South America may decrease by 7 percent.[20]

World wheat production was projected to decline by 3.6 percent from 2011 to 675.1 million tons, with the largest declines in feed and biofuel utilization.[21] The decline can be largely attributed to extreme climatic events.[22] Wheat production in the United States had been expected to increase due to a prediction of more-favorable weather than in 2011, but more than 33 percent of U.S. counties were in severe drought zones—declared natural disaster areas by the U.S. Department of Agriculture.[23]

Since 1960, grain harvest area has increased slightly while production and yield levels have risen dramatically. Grain production has increased 269 percent since 1961, while yield has increased 157 percent.[24] (See Figure 2.) But grain harvest area has only increased 25 percent.[25] (See Figure 3.) The increase of production and yield from 1960 and the significantly smaller increase in grain harvest area are largely due to the Green Revolution and the introduction of high-yielding grain varieties.

Consumption of rice and maize was projected to increase in 2012. Rice consumption per person was expected to reach 57 kilograms.[26] In 2013, rice consumption is expected to drastically change due to India's National Food Security Bill.[27] This program will include subsidized rice that will extend to 75 percent of the rural population and 50 percent of the urban population.[28] The increase in global maize consumption is mainly in feed and industrial uses due to projected increases in meat consumption.[29] Global wheat consumption per person is expected to remain relatively stable.[30] Of the 475.5 million tons of wheat consumed, per capita consumption is about 60 kilograms in developing countries and 97.5 kilograms in industrial nations.[31]

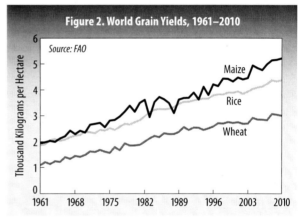

Figure 2. World Grain Yields, 1961–2010

Source: FAO

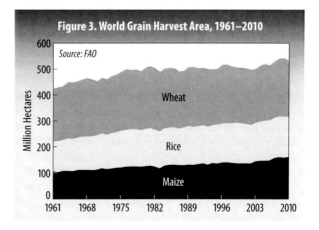

Figure 3. World Grain Harvest Area, 1961–2010

Source: FAO

The reliance on grain crops for food security is threatened by more-extreme climatic events, especially droughts and floods. According to the United Nations International Strategy for Disaster Risk Reduction, the World Food Programme, and Oxfam International, some 375 million people will be affected by climate change–related disasters by 2015.[32] By 2050, FAO notes, 10–20 percent more people will be subject to hunger based on the changing climate's effects on agriculture and 24 million more children are expected to be malnourished—21 percent more than if there were no climate change.[33]

In response to the detrimental effects of climate change on grain crops, there are initiatives to reduce price volatility, move away from fossil-fuel-based agriculture, and recognize the importance of women farmers to increase resilience to climate change. Examples include building up grain reserves, diversifying cropping systems, encouraging agroecology, and supporting women's empowerment in sustainable agriculture.[34] According to the Institute for Trade and Agriculture Policy, "grain reserves are a relatively cheap public insurance policy in the face of tremendous uncertainty, when the risks of failure include starvation."[35]

The relationship between food security, grain production, and climate change was especially pertinent in 2012. The drought taking place in the Midwest and Great Plains of the United States was considered the worst drought in 50 years, coming close to matching the late 1930s Dust Bowl.[36] The drought was expected to cost many billions of dollars and could top the list as one of the most expensive weather-related disasters in U.S. history.[37] The global market will be most affected by this drought, as so much of the developing world relies on U.S. corn and soybean production.[38] Food prices have already begun to increase due to lower yields, and price fluctuations will inevitably affect food security around the globe, especially in the United States and developing countries.[39]

Farmers are also finding ways to build resilience to climate change, including using cover crops, agroforestry, rainwater harvesting, and other agroecological approaches. Agriculture experts at the Rockefeller Foundation have developed a comprehensive list of necessary actions for climate-resilient development.[40] To ensure the ability of farmers to continue to grow crops despite a changing climate, they suggest increased capacity building of agricultural extension agents to reach the most vulnerable communities; partnerships among agricultural institutions, nongovernmental groups, governments, climate organizations, and donors to find relevant solutions; and institutional policy reform that supports the world's most vulnerable farmers.

Disease and Drought Curb
Meat Production and Consumption

Laura Reynolds and Danielle Nierenberg

Global meat production rose to 297 million tons in 2011, an increase of 0.8 percent over 2010 production levels.[1] By the end of 2012, meat production was projected to reach 302 million tons, an increase of 1.6 percent over 2011.[2] These are relatively low rates of growth compared with previous years: in 2010, meat production rose by 2.6 percent, and since 2001 production has risen by 20 percent.[3] (See Figure 1.) According to the U.N. Food and Agriculture Organization (FAO), record drought in the American Midwest, disease outbreaks, and rising prices of livestock feed in 2011 and 2012 all contributed to the lower rises in production.[4] Natural disasters in Japan and Pakistan also constrained output and disrupted trade.[5]

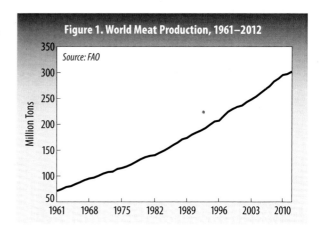

Figure 1. World Meat Production, 1961–2012

Source: FAO

Also bucking a decades-long trend, meat consumption decreased slightly worldwide in 2011, from 42.5 kilograms (kg) per person in 2010 to 42.3 kg.[6] Since 1995, however, per capita meat consumption has increased by 15 percent overall—but consumption in developing countries increased by 25 percent during this time, while in industrial countries it increased by just 2 percent.[7] The rise in consumption was not universal among developing countries; per capita meat consumption in Niger and many other low-income countries remains low.[8] (See Figure 2.) And while the disparity between meat consumption in developing and industrial countries is shrinking, it remains high: the average person in a developing country ate 32.3 kg of meat in 2011, while in industrial countries people on average ate 78.9 kg.[9] Meat consumption was projected to rebound to 2010 levels by the end of 2012, with per capita consumption in industrial countries lowering to 78.4 kg and that in developing countries rising to 32.8 kg.[10]

Pork was the most popular meat in 2011, accounting for 37 percent of both meat production and consumption, at 109 million tons.[11] This was followed closely by poultry meat, with 101 tons produced.[12] Yet pork production decreased by 0.8 percent from 2010, while poultry meat production rose by 3 percent, making it likely that poultry will become the most-produced meat in the next few years.[13] Production of both beef and sheep meat stagnated between 2010 and 2011, at 67 million and 13 million tons, respectively.[14]

A breakdown of meat production by geographic region reveals the dramatic

Laura Reynolds is a staff researcher with the Food and Agriculture Program. **Danielle Nierenberg** formerly directed the Nourishing the Planet Program at Worldwatch Institute.

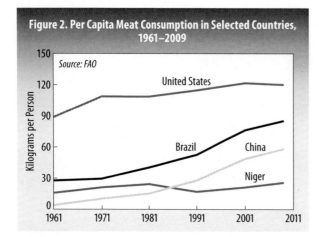

Figure 2. Per Capita Meat Consumption in Selected Countries, 1961–2009

Source: FAO

United States

Brazil

China

Niger

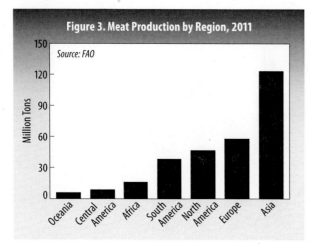

Figure 3. Meat Production by Region, 2011

Source: FAO

shift in centers of production from industrial to developing countries over the last decade. (See Figure 3.) In 2000, for example, North America led the world in beef production, at 13 million tons, while South America produced 12 million tons and Asia, 10 million tons.[15] By 2011, North America had lowered its beef output by 200,000 tons and had been overtaken by both South America and Asia, which produced 15 million and 17 million tons, respectively.[16] FAO attributes the slowdown in growth in industrial countries to rising production costs, stagnating domestic meat consumption, and competition from developing countries.[17] Over the last decade, meat production in Asia grew nearly 26 percent, that in Africa grew 28 percent, and output in South America grew 32 percent.[18]

Widespread and intense drought in China, Russia, the United States, and the Horn of Africa contributed to lower meat production—and to higher meat prices—in 2010 and 2011. Severe drought, intense heat waves, and destructive wildfires in 2010 caused Russia to ban wheat exports.[19] This exacerbated already-high livestock feed and meat prices, and the global effects of Russia's drought were felt well into 2011. At the beginning of that year, China—the world's largest producer and consumer of wheat—experienced its worst drought in about 60 years.[20] This forced the country to increase its wheat imports, driving up grain and livestock feed prices worldwide.[21] Drought conditions in 2011 also affected the Horn of Africa, the continent's largest cattle-producing region, and deterioration of forage in Ethiopia, Kenya, and Somalia caused poor animal conditions and high mortality rates.[22]

In the United States, Texas—the leading cattle-producing state—experienced its worst drought in recorded history.[23] As a result, Texas reported agricultural losses of a record $5.2 billion, including $2.06 billion in livestock losses alone.[24] The United States experienced moderate to extreme drought on 29 percent of its land in summer 2011, including many significant agricultural and grazing areas, bringing corn production to its lowest level in three years and driving up feed and beef prices.[25] According to FAO, the 2011 drought left U.S. cattle herds at their lowest level since 1950.[26] Drought and corn crop failures continued throughout the United States in 2012, causing the U.S. Department of Agriculture to estimate that by 2013 beef would cost 4–5 percent more than in 2010, pork 2.5–3.5 percent more, and poultry 3–4 percent more.[27] The drought and limited livestock numbers in other major exporting countries kept international meat prices at near-record levels in the first quarter of 2012.[28] (See Figure 4.)

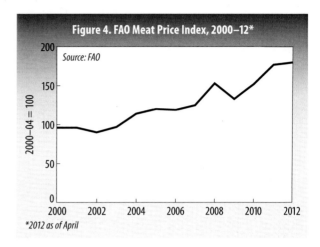

Figure 4. FAO Meat Price Index, 2000–12*

Source: FAO

*2012 as of April

The combination of high prices for meat products and outbreaks of new and recurring zoonotic diseases in 2011 curtailed global meat consumption. Zoonotic diseases, or zoonoses, are diseases that are transmitted between animals and humans. In 2011 alone, foot-and-mouth disease was detected in Paraguay, African swine fever in Russia, classical swine fever in Mexico, and avian influenza (H5N1) throughout Asia.[29] According to a 2012 report by the International Livestock Research Institute, zoonoses cause around 2.7 million human deaths each year, and approximately 75 percent of all emerging infectious diseases now originate in animals or animal products.[30]

Many zoonotic disease outbreaks can be traced to concentrated animal feeding operations (CAFOs), also known as factory farms. These systems now account for 72 percent of poultry production, 43 percent of egg production, and 55 percent of pork production worldwide.[31] And although factory farms originated in Europe and North America, they are becoming increasingly prevalent in developing countries.[32] These systems contribute to disease outbreaks in several ways: they keep animals in cramped and often unsanitary quarters, providing a breeding ground for diseases; they feed animals grain-heavy diets that lack the nutrients needed to fight off disease and illness; and many CAFOs feed animals antibiotics as a preventative rather than a therapeutic measure, causing the animals—and the humans who consume them—to develop resistance to antibiotics.[33]

The most recent figures from the U.S. Centers for Disease Control and Prevention (CDC) suggest that the national frequency of foodborne illness outbreaks has not improved over the past decade. According to the CDC, an estimated 48 million Americans became sick in 2011 from foodborne pathogens, of whom 128,000 were hospitalized and 3,000 died.[34] The most recent statistics from the CDC report that disease outbreaks involving *salmonella*, *vibrio*, *campylobacter*, and *listeria* have all remained steady or increased in prevalence since 2007.[35] Only incidences of *E. coli* have declined within this time period and only marginally so.[36] As the centers of meat production shift from industrial countries to developing ones, and as the methods of meat production become increasingly mechanized and concentrated, governments and corporations must face the real threat that zoonotic diseases are present within the food system.

But not all livestock are reared in industrial or mechanized environments. Nearly 1 billion people living on less than $2 a day depend to some extent on livestock, and many of these people are raising animals in the same ways that their ancestors did.[37] Producing livestock—and their feedgrains—through environmentally sustainable practices can alleviate many of the pitfalls of meat production, including disease outbreaks and susceptibility to drought. Raising native breeds instead of commercial ones—the Milking Shorthorn over the Holstein cow, for instance—may cause slight drops in productivity, but these breeds are also generally less susceptible to heat waves, drought, disease, and native pests.[38] Integrating livestock into farming systems by using manure as fertilizer or by grazing livestock on temporarily fallow fields boosts livestock health, soil fertility, and the farmer's profit. And producing livestock within a local food system can help prevent those diseases that manifest during transport or within industrial slaughter facilities.

Lowering individual meat consumption would also alleviate the pressure to produce more and more meat for lower and lower prices, using rapidly dwindling natural resources. Reconnecting meat production to the land and its natural carrying capacity, as well as reducing meat consumption, can thus greatly improve both public and environmental health.

Farm Animal Populations Continue to Grow

Danielle Nierenberg and Laura Reynolds

Farm animal populations continue to increase worldwide. The number of chickens raised for human consumption increased 169 percent between 1980 and 2010, from 7.2 billion to 19.4 billion.[1] (See Table 1.) During the same period, the population of goats and sheep reached 2 billion, and the cattle population grew 17 percent to reach 1.4 billion.[2] The Consultative Group on International Agricultural Research estimates that by 2050 the global poultry population will grow to nearly 35 billion, the goat and sheep population to 2.7 billion, and the cattle population to 2.6 billion animals.[3]

Demand for meat, eggs, and dairy products in developing countries has increased at a staggering rate in recent decades. (See Figure 1.) According to the U.N. Food and Agriculture and Organization (FAO), between 1980 and 2005 per capita milk consumption in developing countries almost doubled, meat consumption more than tripled, and egg consumption increased fivefold.[4]

The greatest increases in consumption are occurring in East and Southeast Asia. China's per capita milk consumption increased from 2.3 kilograms (kg) in 1980 to 23.2 kg in 2005.[5] Per capita meat consumption in China quadrupled during that period, and egg consumption rose from 2.5 kg to 20.2 kg.[6] And in India, Operation Flood—a National Dairy Board project aimed at boosting the country's milk production and consumption—helped increase per capita milk consumption from 38 kg in 1980 to nearly 69 kg in 2007.[7] India is now the largest milk producer in the world.[8]

People in industrial countries, however, continue to consume the most animal products. Per capita consumption of meat is roughly 78 kg per year in rich countries, while in developing countries people consume just over 32 kg of meat—although that is up from just 15 kg in 1982.[9] Because of growing awareness of the negative health consequences of diets high in animal fat and meat—which include

Table 1. Farm Animal Populations, 1980–2010			
Species	**1980**	**2010**	**Difference**
	(million)		(percent)
Buffalo	121	194	60
Camels	18	24	33
Goats	464	921	98
Pigs	797	965	21
Sheep	1,098	1,078	−1
Ducks	351	1,187	238
Rabbits	194	769	296
Turkeys	313	449	435
Geese	69	359	420
	(billion)		(percent)
Chickens	7.2	19.4	169
Cattle	1.2	1.4	17
	(billion)		(percent)
Total	11.8	26.7	126

Source: FAO, FAOSTAT Statistical Database, at www.faostat.fao.org, updated 23 February 2012.

Danielle Nierenberg formerly directed the Nourishing the Planet Program at Worldwatch Institute. **Laura Reynolds** is a staff researcher with the Food and Agriculture Program.

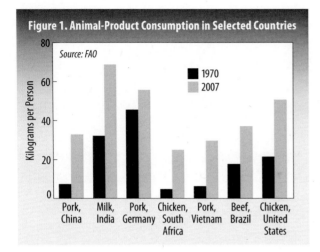

Figure 1. Animal-Product Consumption in Selected Countries

obesity, diabetes, cardiovascular disease, and certain types of cancer—consumption of animal products in many industrial countries, particularly in the European Union, is stagnating or even declining.[10]

The increase in consumption in developing countries can be partly attributed to rising incomes and growing urbanization. According to FAO, urban growth and increases in income are driving higher demand for animal products, particularly in these countries.[11] Currently, more than half of the world's population lives in cities.[12] And between 2010 and 2050, the urban populations of Asia, Africa, and Latin America and the Caribbean are projected to grow to over 5.2 billion.[13] Between 2000 and 2010, the per capita gross national incomes in China and India—two of the world's most rapidly urbanizing countries—increased by 27.8 percent and 51.1 percent, respectively.[14]

Much of the growth in meat, egg, and dairy production is now coming from concentrated animal feeding operations (CAFOs), also known as factory farms. These systems account for 72 percent of poultry production, 43 percent of egg production, and 55 percent of pork production worldwide.[15]

Between 2001 and 2003, these industrial animal operations produced approximately 52.8 million tons of pork worldwide, roughly 10 million tons more than all other pork production systems combined.[16] And poultry operations produced 2.5 times the amount of poultry as all other systems combined.[17] In the United States, milk output per cow has dramatically increased over the last 50 years: in 1961, the average dairy cow produced 7,290 pounds of milk throughout the year; in 2011, the average cow in a factory farm produced 21,335 pounds of milk annually—nearly three times as much.[18]

Unfortunately, factory farms can create serious environmental, animal welfare, public health, and social justice problems. CAFOs produce large amounts of waste, use huge quantities of water and land for feed production, can contribute to the spread of human and animal diseases, and cause biodiversity loss. In addition, diets high in animal fat and meat—particularly red meat and processed meats, such as hot dogs, bacon, and sausage—have been linked to obesity, diabetes, cardiovascular disease, and certain types of cancer.[19]

Because factory farms produce so many animals, the surrounding land cannot absorb their manure, and excess manure is often collected in lagoons or open pits. These produce a smell that is hard to forget, and residents living near CAFOs report more-frequent occurrences of headache, excessive coughing, respiratory problems, nausea, weakness, and burning eyes.[20] Flies and mosquitoes are drawn to the stores of manure—the Ohio Department of Health found, for example, that areas near CAFOs had 83 times more houseflies on average than control areas.[21] Excess manure also frequently spills into nearby waterways, creating huge dead zones or areas of depleted oxygen that fish and other marine life cannot survive in: over the past five years, the dead zone in the Gulf of Mexico has averaged 6,688 square miles—nearly the size of Connecticut and Rhode Island combined.[22]

Due in part to CAFOs' demand for grains like soy and maize, the livestock sector is the world's largest user of Earth's land resources.[23] Livestock grazing occupies 26 percent of Earth's ice-free land surface, and the production of livestock feed uses 33 percent of agricultural cropland.[24] In addition, livestock production is a major driver of deforestation—cattle enterprises have been responsible for 65–80 percent of the deforestation of the Amazon—and countries in South America are clearing large swaths of forest and other land to grow feed crops like maize and soybean.[25]

Livestock also contribute to climate change. Farm animals are responsible for 18 percent of the world's greenhouse gas (GHG) emissions, including 9 percent of carbon dioxide, nearly 40 percent of methane (a GHG 25 times more potent than carbon dioxide), and 65 percent of nitrous oxide.[26]

CAFOs can also be a source of disease among animals and humans. Approximately 75 percent of all emerging diseases originate in animals or animal products.[27] Industrial animal operations can contribute to the spread of swine influenza (H1N1), avian influenza (H5N1), foot-and-mouth disease, mad cow disease, and other diseases.[28] Avian influenza can be particularly contagious and fatal in humans: the World Health Organization has reported nearly 600 human cases of avian flu since 2003, leading to 350 deaths.[29] Enterohaemorrhagic *Escherichia coli*, or *E. coli*, also often originates in factory farms as a result of unsanitary conditions—in 2011 an outbreak of *E. coli* sickened 4,000 people and killed 50 of them across Europe and North America.[30] Because industrial operations are increasingly located close to urban centers, they pose a serious threat to human health.[31]

Widespread sub-therapeutic use of antibiotics in livestock causes the animals—and the humans who consume them—to develop resistance to antibiotics. According to Dr. David Wallinga of the Institute for Agriculture and Trade Policy, "we're sacrificing a future where antibiotics will work for treating sick people by squandering them today [on] animals that are not sick at all."[32]

And as the global livestock population increases, its diversity declines. Industrial meat operations rely on a narrow range of commercial breeds selected for their high productivity—two cow breeds, Holstein and Jersey, make up 97 percent of the U.S. dairy-cow herd.[33] As a result, indigenous livestock breeds, which have evolved to the specific climate, terrestrial, and disease characteristics of their regions, are rapidly disappearing: in 2010, FAO reported that at least 21 percent of the world's livestock breeds are at risk of extinction.[34] It estimated that between 2002 and 2007, one breed of cattle, goats, pigs, horses, or poultry was lost every month on average.[35]

This narrowing of genetic diversity greatly compromises livestock producers' ability to withstand the challenges of climate change, including water supply changes, lack of forage, disease expansion, and increasing temperature variation. Imported commercial breeds usually do not carry resistance to these stresses, putting them at greater risk for disease, starvation, and heat exhaustion. Jacob Wanyama, coordinator for the African LIFE Network, an advocacy group for pastoralist communities in East Africa, says that indigenous breeds like the East African Ankole cattle are not only "beautiful to look at" but can also survive in extremely harsh, dry conditions and do not require expensive feed and inputs, such as antibiotics, to keep them healthy.[36]

Although industrial animal operations originated in Europe and North America, they are becoming increasingly prevalent in developing countries. In East and Southeast Asia, for example, meat production increased by 25 million tons, or 31 percent, between 2001 and 2007 alone, and most of this growth took place in industrial systems.[37] FAO estimates that 80 percent of growth in the livestock sector now comes from industrial production systems.[38] And in many developing regions, environmental, animal welfare, public health, and labor standards are not as well established as in North America and Europe.[39]

Globally, the livestock sector contributes 40 percent of gross global agricultural product and employs 1.3 billion people.[40] Livestock act as living banks, providing farmers and livestock keepers with insurance for loans, investments for the future, or quick cash in times of emergency.[41] In Rwanda, for example, where genocide derailed development and killed 1 million people in 1994, Heifer International has helped farmers in the Gicumbi District in the north increase their incomes and spread self-sufficiency by training them to raise dairy cows and pass on their expertise to other farmers.[42] Livestock raised locally in a sustainable production system can also contribute to gender equality and opportunities for women, improve the structure and fertility of the soil, provide draught power, and control insects and weeds.[43] Manure from livestock can also provide cooking and heating fuel for the estimated 1.3 billion poor or rural families who lack access to electricity.[44]

Policymakers need to find ways to produce meat and other animal products in environmentally and socially sustainable ways. This can be done by establishing stricter regulations, such as waste management and zoning laws, for industrial producers; by giving pastoralists land titles for their traditional grazing sites; by helping farmers gain access to critical financial services, including credit, insurance, and broader markets; and by funding research and education on the benefits of raising indigenous livestock breeds.[45]

Aquaculture Tries to Fill World's Insatiable Appetite for Seafood

Katie Spoden and Danielle Nierenberg

Total global fish production, including both wild capture fish and aquaculture, reached an all-time high of 154 million tons in 2011.[1] (See Figure 1.) Wild capture was 90.4 million tons that year, up 2 percent from 2010.[2] This followed a 1.6 percent decline from 2009 to 2010.[3] The 2011 global capture figure matched the 2007 total of 90.3 million tons, which broke a four-year pattern of declining global wild capture.[4] Since the late 1980s, however, wild capture production has essentially stagnated.[5]

Aquaculture, in contrast, has been expanding steadily for the last 25 years and saw its largest increase in 2010, when it grew 8.7 percent to 59.9 million tons.[6] In 2011, production rose again by 6.2 percent, to 63.6 million tons.[7]

Aquaculture currently provides about 40 percent of the fish consumed globally and is expected to top 60 percent by 2020.[8] Growth in fish farming can be a double-edged sword, however. Despite its potential to affordably feed an ever-growing global population, it can also contribute to problems of habitat destruction, waste disposal, invasions of exotic species and pathogens, and depletion of wild fish stock.[9]

In 2011, inland aquaculture increased 6.2 percent to reach 44.3 million tons, while marine aquaculture increased 6.6 percent, to 19.3 million tons.[10] Inland wild capture totaled 11.5 million tons, an increase of 2.7 percent.[11] Marine capture production grew much less, just 1.9 percent, to a total of 78.9 million tons.[12]

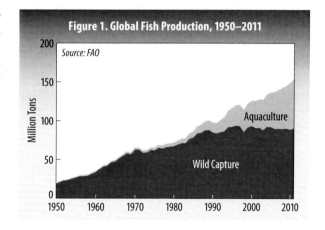

Figure 1. Global Fish Production, 1950–2011

Source: FAO

Nonetheless, this accounted for the largest share of global fish production—51 percent—followed by inland aquaculture at 29 percent.[13] (See Figure 2.)

Fish production rose 6.4 percent in Asia in 2010 (the latest year with regional data), amounting to 121.3 million tons—making this region the dominant producer of both captured and farmed fish at 72 percent of the global total.[14] (See Figure 3.) In 2010, Europe, a distant second, produced 9.7 percent (16.4 million tons) of the global fish supply.[15] Latin America and the Caribbean produced 8.3 percent, Africa 5.4 percent, North America 3.7 percent, and Oceania just 0.8 percent.[16]

Humans ate 130.8 million tons of fish in 2011.[17] The remaining 23.2 million tons of fish went to such non-food uses as fishmeal, fish oil, culture, bait, and phar-

Katie Spoden was a food and agriculture intern at Worldwatch. Danielle Nierenberg formerly directed the Nourishing the Planet Program.

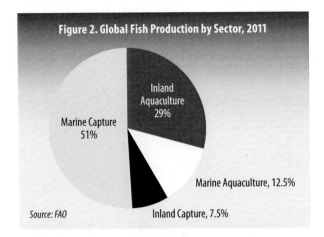

Figure 2. Global Fish Production by Sector, 2011

Inland Aquaculture 29%

Marine Capture 51%

Marine Aquaculture, 12.5%

Inland Capture, 7.5%

Source: FAO

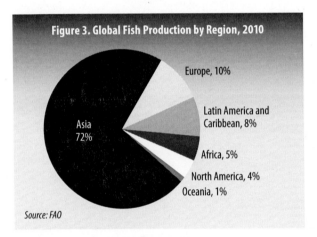

Figure 3. Global Fish Production by Region, 2010

Europe, 10%

Latin America and Caribbean, 8%

Asia 72%

Africa, 5%

North America, 4%

Oceania, 1%

Source: FAO

maceuticals.[18] The human consumption figure increased 14.4 percent in the last five years.[19] And consumption of farmed fish has risen tenfold since 1970, at an annual average of 6.6 percent per year.[20] Asia consumes two thirds of the fish caught or grown for consumption.[21]

The largest increases in consumption are concentrated in East Asia, Southeast Asia, and North Africa.[22] The lowest consumption in 2009 was in Africa, at 9.1 kilograms per person.[23] On average, people in Oceania ate 24.6 kilograms of fish; in North America, 24.1 kilograms; in Europe, 22.0 kilograms; and in Latin America and the Caribbean, 9.9 kilograms.[24]

Fish are an essential source of protein worldwide, providing more than 20 percent of the total animal protein supply for approximately 3 billion people.[25] Chinese per capita consumption of fish reached 31.9 kilograms in 2009, averaging an annual growth rate of 6 percent from 1990 to 2009.[26] Chinese consumers receive 8.2 percent of their total protein from fish.[27] In Japan, the second largest producer of fish, consumers get 21 percent of their protein from fish.[28] In 2010, some 400 million people from the poorest African and South Asian countries relied on fish for more than half their total protein intake.[29]

The fish sector is also a source of income and sustenance for millions of people worldwide. During 2005–10, employment in fisheries and aquaculture increased 2.1 percent annually compared with world population growth of 1.2 percent and growth of jobs in traditional agriculture of 0.5 percent.[30] In 2010, some 54.8 million people were directly engaged full-time or part-time in capture fisheries or aquaculture.[31] More than 87 percent of these individuals lived in Asia, 7 percent in Africa, and just 3.6 percent in Latin America and the Caribbean.[32]

According to the U.N. Food and Agriculture Organization, for every one job in the fish sector, three to four additional jobs are produced in secondary activities, such as fish processing, marketing, maintenance of fishing equipment, and other related industries.[33] And on average each person working in the fish sector is financially responsible for three dependents.[34] In combination, then, jobs in the primary and secondary fish sectors support the livelihoods of 660 million to 820 million people—10 to 12 percent of global population.[35]

Although Africa is only the fourth largest producer of fish in the world, its water resources are highly sought after by larger, more-competitive fishing trawlers. Extreme overfishing occurs when foreign trawlers buy fishing licenses from African countries for marine water use. In West African waters, foreign trawlers pose a threat because factory ships from the United Kingdom, other countries within the European Union, Russia, and Saudi Arabia can outcompete the technologies used by local fishers.[36] In Senegal, for example, a local fisher can catch a few tons of fish each day in the typical 30-foot *pirogue*.[37] In contrast, factory ships from industrial countries catch hundreds of tons daily in their 10,000-ton factory ships.[38]

Wild fish stocks are at a dangerously unsustainable level. As of 2009 (the most recent year with data), 57.4 percent of fisheries were estimated to be fully exploited—meaning current catches were at or close to their maximum sustainable yield, with no room for further expansion.[39] Of the remaining fisheries in jeopardy, 29.9 percent were deemed overexploited, while only 12.7 percent were considered to be not fully exploited.[40]

Ten species of fish account for 30 percent of the world's marine capture.[41] (See Figure 4.) Of these 10, most of the stocks are considered to be either fully exploited (meaning no room for further expansion) or overexploited (meaning with effective rebuilding and sustainable measures, increases in production may be possible).[42] Of the stocks with no room for further expansion, the two main stocks of anchoveta, Alaskan pollock, blue whiting, Atlantic herring, and chub mackerel are considered fully exploited.[43] Largehead hairtail, Japanese anchovy, and Chilean jack mackerel are thought to be overexploited.[44] Tuna is the species most at risk: of the seven principal stocks of tuna, 33 percent are overexploited, 37.5 percent are fully exploited, and only 29 percent are not fully exploited.[45]

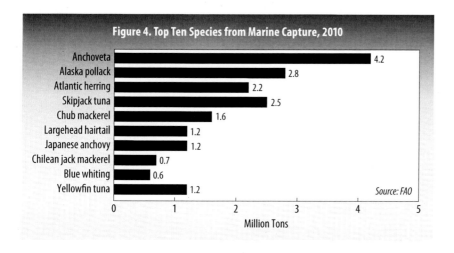

Figure 4. Top Ten Species from Marine Capture, 2010

A number of government initiatives give some hope to a future of sustainable fishing. In the United States, the Magnuson-Stevens Act mandated that overfished stocks be restored; as of 2012, two thirds of U.S. stocks were fished sustainably and only 17 percent were fished at overexploited levels.[46] In New Zealand, 69 percent of stocks were above management targets, but Australia only reports 12 percent of stocks at overexploitation levels due to increased government fishery standards.[47]

To maintain the current level of fish consumption in the world, aquaculture will need to provide an additional 23 million tons of farmed fish by 2020.[48] To produce this additional amount, fish farming will also have to provide the necessary feed to grow the omnivorous and carnivorous fish that people want. Aquaculture is being pressured to provide both food and feed because of the oceans' fully tapped and overexploited fisheries.[49]

Aquaculture is providing a rapidly growing share of fishmeal and fish oil, accounting, respectively, for 61 and 74 percent in 2008.[50] More research and development is needed to find affordable replacements for fishmeal and terrestrial products in aquafeeds. Alternatives include nutrient waste streams produced by other species inhabiting the same aquatic ecosystem, as well as by-products from terrestrial agriculture, including non-food-grade livestock by-products, plant oilseed and legume meals, and cereal by-products. [51]

Continually increasing fish production, from both aquaculture and fisheries, raises many environmental concerns. If aquaculture continues to grow without constraints, it could lead to degradation of land and marine habitats, chemical pollution from fertilizers and antibiotics, the negative impacts of invasive species, and a lessened fish resistance to disease due to close proximity and intensive farming practices.[52]

Just like cattle, fish are vulnerable to disease when they are in close proximity to one another. Disease outbreaks have affected Atlantic salmon in Chile, oysters in Europe, and marine shrimp farming in Asia, South America, and Africa.[53] In 2010, China lost 1.7 million tons of fish due to natural disasters, diseases, and pollution, and in 2011 Mozambique suffered severe losses in marine shrimp farming due to disease.[54] Fish producers often use antibiotics to combat these diseases. According to Chile's Ministry of Economy, Development, and Tourism, almost 718,000 pounds of antibiotics were used in 2008 and more than 850,000 pounds of antibiotics were used in 2007.[55] In March 2010, the Chilean Congress banned the preventative use of antibiotics and demanded that companies make the amount and reason for antibiotic use available to the public.[56]

Rice farmers in China, Indonesia, and Egypt are diversifying farming methods by growing fish in rice-based ecosystems. Almost 90 percent of the world's rice crop provides a suitable environment for growing fish and other aquatic organisms.[57] Fish culture in rice paddies is practiced mainly in Asia but also in a number of countries in other regions.[58] The benefits include additional food and income for farmers who choose to diversify and less money spent on pesticides and fertilizers.[59] The fish feed on rice pests: an integrated rice-fish system uses 68 percent less pesticide than rice monoculture and emits 30 percent less methane.[60] Especially in China, with 15 percent of the suitable rice-fish ecosystems, there is room for growth in this form of fish farming.[61]

Fisheries and aquaculture in the future will be heavily affected by a growing population and increasing fish consumption, by economic pressures on scarce natural resources, and by climate change. Pavan Sukhdev, former head of the United Nations Environment Programme's Green Economy Initiative, predicts "if the various estimates we have received…come true, then we are in the situation where 40 years down the line we, effectively, are out of fish."[62] We will have overfished to a point of no return by 2050.

To prevent these problems, policymakers, fishers, and consumers need to find alternative sources for fish feed, combat illegal fishing, encourage more-sustainable practices in aquaculture, acknowledge the potential effects of climate change on the oceans, and think critically about what and how much fish to consume.

Area Equipped for Irrigation at Record Levels, But Expansion Slows

Judith Renner

In 2009, the most recent year for which global data are available from the U.N. Food and Agriculture Organization (FAO), 311 million hectares in the world were equipped for irrigation.[1] (See Figure 1.) As of 2010, the countries with the largest areas were India (66.3 million), China (62.9 million hectares), and the United States (24.7 million).[2]

Worldwide, 84 percent of the area equipped for irrigation is actually being irrigated.[3] The share is highest in Asia (87 percent) and Africa (85 percent).[4] It is somewhat lower in the Americas (81 percent) and in Oceania (77 percent), but much lower in Europe (59 percent).[5] (See Figure 2.) Higher and more reliable levels of rainfall allow parts of Europe—particularly northern and eastern Europe—to rely less on existing irrigation infrastructure than is the case in drier or more variable climates.

The irrigation sector claims about 70 percent of freshwater withdrawals worldwide.[6] Irrigation can offer crop yields two to four times greater than is possible with rainfed farming.[7] Indeed, irrigated areas provide 40 percent of the world's food from approximately 20 percent of its agricultural land.[8]

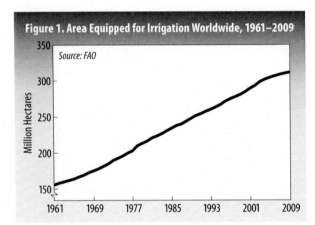

Figure 1. Area Equipped for Irrigation Worldwide, 1961–2009

Source: FAO

Million Hectares

Since the late 1970s, irrigation expansion has experienced a marked slowdown.[9] (See Figure 3.) According to an FAO assessment, unsatisfactory performances of formal large canal systems, corruption involved in the construction process, and acknowledgement of the environmental impact of irrigation projects all contributed to the decrease in investments.[10]

Also, over the past 30 years the shift from public to private investment in irrigation has been driven by the increasing availability of inexpensive individual pumps and well construction methods.[11] Groundwater is generally less prone to pollution than surface water, and the use of aquifers is increasing worldwide.[12] The takeoff in individual groundwater irrigation has been concentrated in India, China, and much of Southeast Asia.[13] The idea of affordable and effective irrigation is an attractive one to poor farmers, with rewards of higher outputs and incomes and better diets.[14]

The option is often made even more appealing with offers of government subsidies for energy costs of running groundwater pumps and support prices of irrigated

Judith Renner is a senior at Fordham University in New York, studying international political economy and sociology.

products.[15] In India's Gujarat state, for example, energy subsidies are structured so that farmers pay a flat rate, no matter how much electricity they use.[16] But with rising numbers of farmers tapping groundwater resources, more and more aquifers are in danger of overuse.[17]

If groundwater resources are overexploited, aquifers will be unable to recharge fast enough to keep pace with water withdrawals. It should be noted that not all aquifers are being pumped at unsustainable levels—in fact, 80 percent of aquifers worldwide could handle additional water withdrawals.[18] One troubling aspect of groundwater withdrawals is that the world's major agricultural producers (particularly India, China, and the United States) are also the ones responsible for the highest levels of depletion.[19]

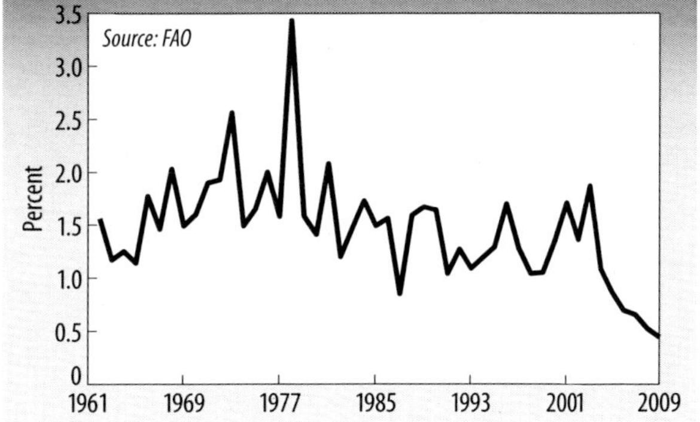

Another problem with pumping water from aquifers and redirecting flows for irrigation is the impact on delicate environmental balances. Salinization occurs when water moves past plant roots to the water table due to inefficient irrigation and drainage systems; as the water table rises, it brings salts to the base of plant roots.[20] Plants take in the water, and the salts are left behind, degrading soil quality and therefore the potential for growth. When plants become waterlogged from over-irrigation, these effects of salinity are worsened. Plant roots take in the salts along with

Figure 2. Irrigated Land by Region and Percent Actually Irrigated

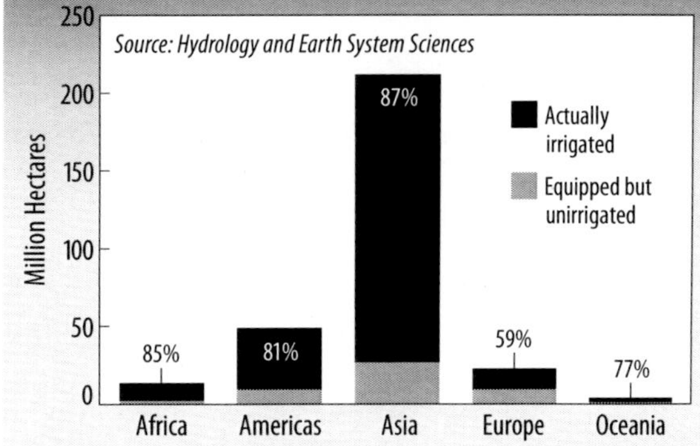

Source: Hydrology and Earth System Sciences

Actually irrigated

Equipped but unirrigated

Figure 3. Annual Increase in Area Equipped for Irrigation, 1962–2009

Source: FAO

the excess water, which could further stunt growth.[21] Salinization and waterlogging also reduce potential nesting sites for birds, food sources for wildlife, and biodiversity in streams and wetlands.[22] Salinity decreases food production and income, which presents a concern for those who rely on agriculture for their livelihoods and for those who buy their products.[23]

Some countries have started looking abroad for places to grow food in the face of unmet irrigation needs. Land grabbing, predominantly in Africa, has become a prominent trend. This involves selling or leasing agricultural land to private and public investors abroad for the purpose of growing and exporting food. For example, in recent years Saudi Arabian companies have been buying up millions of hectares of land in Ethiopia and elsewhere in Africa to guarantee food security at home.[24] Foreign companies also buy up land with the knowledge that they will have free access to the precious water resources contained within the purchased area.[25] As with Saudi Arabia, the underlying reason that many companies engage in African land grabs is the opportunity for a "water grab."[26]

With a rising population demanding greater agricultural output and with water resources that are steadily declining, it is crucial to use irrigation methods that yield "more crop per drop."[27] Flood irrigation remains most commonly used by farmers, even though plants often use only about half the amount of water applied in that system.[28] In some cases, the excess water returns to rivers or groundwater aquifers, where it can be reused, but too much water use can rapidly deplete resources or counteract growth through salinization.[29]

A potentially better alternative is drip irrigation, a form of micro-irrigation that waters plants slowly and in small amounts either on the soil surface or directly on roots.[30] These techniques have the potential to reduce water use by as much as 70 percent while increasing output by 20–90 percent.[31] Drip irrigation was first used on a large scale in the 1970s in Australia, Israel, Mexico, New Zealand, South Africa, and the United States to produce fruits and vegetables.[32] Within the last two decades the area irrigated using drip and other micro-irrigation methods has increased 6.4-fold, from 1.6 million hectares to over 10.3 million hectares.[33] India currently claims the lead worldwide, irrigating almost 2 million hectares of its land using these micro methods.[34]

In June 2012, David Hillel, an Israeli scientist who pioneered an innovative way of efficiently delivering water to crops in arid and dryland regions, was awarded the 2012 World Food Prize—an indication that more-efficient forms of irrigation are receiving growing attention.[35] Hillel developed methods to apply water in small amounts directly to plant roots, thus dramatically cutting the amount of water required to nourish crops and permitting higher crop yields.[36]

While increasing production and profits, drip irrigation is also a more environmentally sound choice as it reduces leaching and excess runoff.[37] A drawback is the expense attached to this method of irrigation; as drip systems require a larger investment than flood irrigation, many poor farmers cannot afford them.[38] Also, there is no great incentive for investing in water-saving methods in countries like India that provide government subsidies for both water and energy costs.[39] Within the last 10 years, companies like iDE (formerly International Development Enterprises) have begun to develop and market inexpensive drip systems to low-income farmers, and more than 600,000 of these have been sold in India, Nepal, Zambia, and Zimbabwe.[40]

Climate change represents a serious concern for irrigation prospects. By 2080, the atmospheric temperature is predicted to rise by about 4 degrees Celsius.[41] Shifts in the hydrological cycle are expected as a result of this global warming, causing rainfall and temperatures to be more inconsistent.[42] Demand for water will continue to rise, since even small decreases in rainfall will result in substantial reductions in valuable runoff.[43] Rising sea levels will also take a toll on agriculture by affecting drainage and water levels in coastal regions, which could cause salinization of aquifers and river estuaries.[44] These consequences of climate change will contribute to decreasing agricultural productivity and heavy demand for water.

With predictions of a global population of over 9 billion by 2050, demand for higher agricultural output will put more strain on already fragile water reserves.[45] Even without the effects of climate change, water withdrawals for irrigation will need to rise by 11 percent in the next three decades to meet crop production demands.[46] Reconciling increasing food demands with decreasing water security requires efficient systems that produce more food with less water and that minimize water waste. Intelligent water management is especially crucial in the face of climate change, which will force the agriculture industry to compete with the environment for water.

Organic Agriculture Contributes to Sustainable Food Security

Catherine Ward and Laura Reynolds

In 2010, the most recent year for which data are available, organic farming accounted for approximately 0.9 percent of total agricultural land around the world.[1] While this is still a minuscule share, since 1999 the land area farmed organically has expanded more than threefold: 37 million hectares of land are now organically farmed, including land that is in the process of being converted from conventional agricultural practices.[2] (See Figure 1.)

The amount of organically farmed land dropped very slightly, by 0.1 percent, between 2009 and 2010.[3] A decline in this land in India and China was almost matched by an increase in Europe. Regions with the largest organic agricultural land in 2010 were Oceania, including Australia, New Zealand, and Pacific Island nations (12.1 million hectares); Europe (10 million hectares); and Latin America (8.4 million hectares).[4] (See Figure 2.)

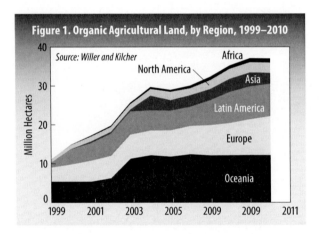

Figure 1. Organic Agricultural Land, by Region, 1999–2010

Organic farming is now established in international standards, and 84 countries had implemented organic regulations by 2010, up from 74 countries in 2009.[5] Definitions vary, but according to the International Federation of Organic Agriculture Movements, organic agriculture is a production system that relies on ecological processes rather than the use of synthetic inputs, such as chemical fertilizers and pesticides.[6]

The modern organic farming movement emerged in the 1950s and 1960s, largely as a reaction to consumer concerns over the use of agrochemicals.[7] The period after World War II and through the 1950s is commonly known as the "golden age of pesticides" because the use of agricultural chemicals was widespread and their effects were largely unknown.[8] As the health and ecological consequences of agrochemicals began to be understood, governments started to regulate their use and consumers started to demand organically certified foods.[9]

While organic agriculture often produces lower yields on land that has recently been farmed conventionally, it can outperform conventional practices—especially in times of drought—when the land has been farmed organically for a longer time.[10] But increased yields alone are not meeting the needs of people around the world.[11] Globally, at least 1 billion people do not have adequate access to sufficient food—due to issues of distribution, individual purchasing power, storage

Catherine Ward was a research intern and **Laura Reynolds** is a staff researcher with the Food and Agriculture Program at Worldwatch Institute.

and refrigeration, and market access, among others—with sub-Saharan Africa being the region most affected.[12] Conventional agricultural practices often degrade the environment over both the long and the short term through soil erosion, excessive water extraction, and biodiversity loss.[13]

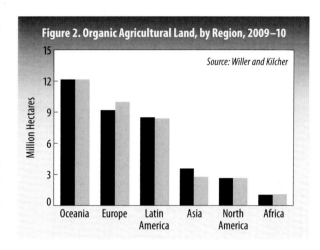

Figure 2. Organic Agricultural Land, by Region, 2009–10

Source: Willer and Kilcher

Organic farming has the potential to contribute to sustainable food security by improving nutrition intake and sustaining livelihoods in rural areas, while simultaneously reducing vulnerability to climate change and enhancing biodiversity.[14] Sustainable agricultural practices associated with organic farming are relatively labor-intensive and have the potential to contribute to long-term employment in rural areas.[15] Organic agriculture uses up to 50 percent less fossil fuel energy than conventional farming, and common organic practices—including rotating crops, applying mulch to empty fields, and maintaining perennial shrubs and trees on farms—also stabilize soils and improve water retention, thus reducing vulnerability to harsh weather patterns.[16] On average, organic farms have 30 percent higher biodiversity, including birds, insects, and plants, than conventional farms do.[17]

Certifications for organic agriculture are increasingly concentrated in wealthier countries.[18] From 2009 to 2010, Europe increased its organic agricultural land share by 9 percent to 10 million hectares, the largest regional growth of land in this category.[19] Despite rapid expansion of certified organic agriculture over the last decade, the United States has lagged behind other countries in adopting these farming methods.[20] When national sales rather than production are considered, however, the U.S. organic industry is one of the fastest-growing industries in the nation, expanding by 9.5 percent in 2011 to reach $31.5 billion in sales.[21]

Sustainable food production will become increasingly important in developing countries, as the majority of population growth is concentrated in the world's poorest countries.[22] Agriculture in developing countries is often far more labor-intensive than in industrial countries, and therefore it is not surprising that approximately 80 percent of the 1.6 million global organic farmers live in developing countries.[23] The countries with the most certified organic producers in 2010 were India (400,551 farmers), Uganda (188,625 farmers), and Mexico (128,826 farmers).[24]

Africa accounts for 3 percent of organic agricultural land in the world, with just over 1 million hectares of certified organic land.[25] (See Table 1.) Organic farming in Africa is now being recognized as a way to address problems of food insecurity and climate change.[26] Small plot farming in Zambia, Malawi, Niger, and Burkina Faso has drawn on organic methods to restore soils; this has resulted in higher food crop yields, greater household food security, and increased incomes.[27] A combination of traditional and organic farming techniques—involving water harvesting, composting, and applying mulch to the land—has allowed farmers in Burkina Faso to adapt to climate change and build resilience to weather shocks.[28] In Ethiopia, organic

Table 1. Global Organic Land Distribution, by Region, 2010

Region	Organic Area	Share of Global Organic Land
	(million hectares)	(percent)
Africa	1.07	3
North America	2.65	7
Asia	2.78	7
Latin America	8.39	23
Europe	10.00	27
Oceania	12.14	33
Total	37.03	100

Source: Helga Willer and Lukas Kilcher, eds., The World of Organic Agriculture—Statistics and Emerging Trends 2012 *(Bonn and Frick: IFOAM and FiBL, 2012).*

farming methods have helped farmers use water more efficiently and restore soil health to better withstand harsh weather conditions like drought, while increasing crop yields and improving food security.[29]

Asia has 7 percent of the world's organic agricultural land, with a total of 2.8 million hectares.[30] Despite a decline in organically farmed land in China and India between 2009 and 2010, India's export volume of organic produce increased by 20 percent.[31] Small-scale farmers in India, who account for at least 70 percent of the nation's farming community, are reluctant to engage in organic farming due to problems getting enough organic supplements, lack of access to certification, and limited local market access for organic produce.[32] In Cambodia, on the other hand, negative impacts of conventional farming systems on the environment and on farmers have resulted in widespread conversion to organic agriculture.[33] Health indicators, such as pesticide poisoning–related symptoms, improved among Cambodian farmers who adopted organic techniques.[34]

The global food system will experience greater pressure in the decades ahead to produce more food to meet the demands of a growing population.[35] Increasing food production alone is not sufficient to combat hunger, particularly among small-scale farmers in developing countries. Over 70 percent of the world's poor live in rural areas and depend directly on agriculture for their income.[36] Small-scale farmers face a number of constraints in adopting organic farming practices—practices that should be integrated with local farmers' needs and knowledge systems into national frameworks that are supported by government agencies and non-governmental organizations.[37]

Investing in Women Farmers

Seyyada A. Burney and Danielle Nierenberg

Women farmers produce more than half of all food worldwide and currently account for 43 percent of the global agricultural labor force.[1] Indeed, the global food and agriculture system depends more on the contributions of women farmers today than ever before. Women produce as much as 50 percent of the agricultural output in South Asia and 80 percent in sub-Saharan Africa.[2] (See Table 1.)

Women's labor varies by country, ethnicity, and type of farming. In Indonesia, women generally undertake labor-intensive, low value-added rice paddy farming while men typically take part in more-profitable and mechanized rubber plantation work.[3] In the United States, women represent a major proportion of the growing organic and smallholder farming trend, while men are more frequently involved in large-scale agribusiness.[4] In countries throughout Africa, Asia, and the Pacific, women also work an additional 12–14 hours per week in invisible or unquantifiable roles such as providing child care, collecting firewood, preserving biodiversity, or cooking.[5]

In spite of women farmers' essential roles in global and local food security, there is a persistent gender gap in agriculture.[6] Women represent 70 percent of the 1.3 billion people living in poverty around the world.[7] Despite high levels of agricultural productivity, persistent inequity limits women's full participation in local economies.[8] Cultural norms and restrictive property or inheritance rights limit the types and amount of financial resources, land, or activity available to women. Studies in South Asia and throughout the Middle East also show that women receive lower wages and are more likely to work part-time or seasonally than men in comparable jobs, regardless of similar levels of education and experience—in Bangladesh, for example, only 3 percent of rural women take part in paid employment, compared with 24 percent of rural men.[9]

Recognizing the factors restricting women from receiving full compensation for their role in global agriculture is key to alleviating the gender gap in agricultural employment, resources, and development. Women produce 60–80 percent of the food in developing countries but own less than 2 percent of the land.[10] They typically farm non-commercial, staple crops, such as rice, wheat, and maize, which account for 90 percent of the food consumed by the rural poor.[11]

Fewer extension or research services are directed at women farmers because of perceptions of the limited commercial viability of their labor or products—and only 15 percent of extension officers around the world are women.[12] Yet the Economist Intelligence Unit's newly developed Global Food Security Index has a 0.93 correlation with its index of Women's Economic Opportunity, showing that countries with more gender-sensitive business environments, based on labor policies,

Seyyada A. Burney was a research intern with Nourishing the Planet at Worldwatch Institute. Danielle Nierenberg formerly directed the Nourishing the Planet Program.

Table 1. Women in Agriculture in Developing Countries	
Sub-Saharan Africa	Cultural norms encourage independence and self-reliance among women. Women produce up to 80 percent of agricultural output in this region, though individual asset ownership is limited. Conflict, gendered migration patterns, and HIV/AIDS have led to increased reliance on women's contributions to food security and farming. Women farmers currently represent 36 percent of the agricultural labor force in Côte d'Ivoire and Niger and up to 60 percent in Lesotho.
East and Southeast Asia	Regional statistics for female participation in agriculture are dominated by China, where 48 percent of farmers are women.
South Asia	Recent food crises have led to rapid increases in women's agricultural labor. Women farmers currently represent 30 percent of the agricultural labor force in India and 50 percent in Bangladesh. Female participation in Pakistan's agriculture has tripled since the 1980s.
Near East and North Africa	The percentage of women farmers in the agricultural labor force has increased 15 percent since 1980. However, only 15 percent of landholders are women.
Latin America	Overall, women's participation in agriculture is high, but women represent a relatively low proportion of the labor force as a result of higher education levels and greater social and economic mobility. Regional averages for female farmers as a percentage of the labor force are steady at 20 percent.

Source: FAO, 2010–2011 The State of Food and Agriculture—Women in Agriculture *(Rome: 2011).*

access to finance, and comparative levels of education and training, have more abundant, nutritious, and affordable food.[13] This relationship provides irrefutable evidence that when women have equal resources and opportunity they can produce higher—and higher-quality—agricultural yields.[14]

Low levels of education limit women's ability to adapt to new agricultural technology, navigate market fluctuations, and make decisions within their households and communities. Over two thirds of the world's 1 billion illiterate adults are women.[15] Local illiteracy rates can be high in countries such as Burkina Faso, where 78 percent of rural women cannot read or write.[16] Economic necessity and cultural practices cause many girls to drop out of school in order to marry or take care of invalid or elderly family members—two thirds of the 130 million children missing from schools are young girls.[17] One in five women worldwide give birth before the age of 18—many of them concentrated in rural areas of middle- to low-income countries, where this number can rise to one in three.[18]

Research shows that each additional year of primary education can increase a girl's eventual income by 10–20 percent, while an extra year of secondary school can raise earnings by 12–25 percent.[19] The Punjab Education Sector Reform Program in Pakistan increased female school enrollment by 11 percent in addition to supplementing family income in poor households by providing a stipend to girls aged 10–14 so they could attend school.[20]

Studies in Bangladesh also connect the "Asian Enigma"—lagging child nutrition despite substantial economic and agricultural productivity gains—to women's empowerment by showing that long-term nutritional status is higher in households where women are better educated.[21] A Nigerian study reaffirms the importance of maternal health and education by tracing child malnutrition back to low birth weight, a key indicator of maternal age and nutrition, and to low levels of female secondary school enrollment.[22] The health of women and young girls is more vulnerable to fluctuations in family income than boys' health because social norms dictate that women and girls often eat less in times of financial hardship, leading to greater incidences of female undernourishment and to nutrition problems such as iron deficiency, which in turn limits their labor productivity and that of their children.[23]

Time constraints, cultural practices, and historic gender biases discourage women farmers from making use of technology and extension services designed to improve agricultural yields. Disparities in the use of mechanical technology between rural men and women exist in at least 13 countries (see Figure 1), a trend that can be attributed to many different factors: lower levels of education, credit, or extension services among women and gender biases among technology distributers or users.[24]

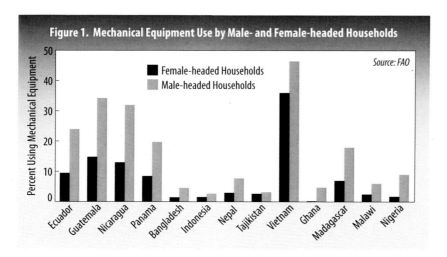

Figure 1. Mechanical Equipment Use by Male- and Female-headed Households

Female-headed households are less able to respond to food crises or fluctuating commodity prices by increasing production. Women farmers in sub-Saharan Africa, for instance, face substantial yield losses and cannot participate in a second cropping season because of plowing or planting delays caused by poor technology and a lack of financial resources.[25] Only 10–20 percent of landholders are women in most developing countries (see Figure 2)—and as few as 5 percent in the Middle East and North Africa—and farms run by female-headed households are typically one half to two thirds the size of those handled by male-headed households.[26]

Women typically own smaller and lower-quality or less-profitable assets, such as poultry rather than cattle; men's livestock holdings are up to three times larger in Bangladesh, Ghana, and Nigeria, for example.[27] Smallholder credit use is 5–10 percentage points lower among women as a direct result of limited assets such

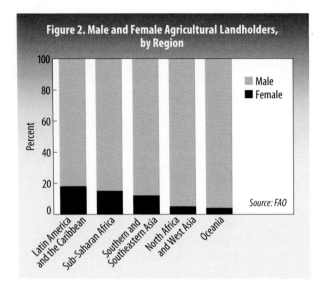

Figure 2. Male and Female Agricultural Landholders, by Region

Source: FAO

as livestock or property for collateral.[28] In addition, there are direct correlations between women's asset ownership and family nutritional outcomes: the Organisation for Economic Co-operation and Development (OECD) found that countries where women's property rights are restricted have, on average, 60 percent more cases of child malnourishment than countries with more-inclusive property rights or credit opportunities.[29]

Farmers in countries with greater gender equality, based on an OECD Index of Social Institutions and Gender Inequality, tend to achieve higher average cereal yields than those in countries with more inequality.[30] The countries are also more food-secure, based on food affordability, availability, quality, and safety.[31] Improved agricultural productivity reinforces gains in gender equality in addition to creating a positive feedback mechanism throughout local communities.

Women reinvest up to 90 percent of their income in child and household well-being whereas men reinvest only 30–40 percent.[32] In Ethiopia, a project run by the International Center for Research on Women found that the incomes of female-headed households increased by 18 percent, or $268, when the women used new treadle irrigation pumps, while male-headed households saw only a 14 percent rise in income using the same technology.[33] Using renewable fuel sources, such as solar energy, reduces household expenditures and the amount of time rural women spend collecting fuel.[34]

When provided with equitable resources, women farmers can generate substantial gains in agricultural productivity. For instance, increased individual smallholder yields as a result of closing gender gaps in land ownership can raise domestic agricultural output by 2.5–4 percent.[35] This level of improved food production could reduce domestic undernourishment cases by 12–17 percent, a direct impact that could be even greater in countries where there is a more pronounced gender gap.[36] Over time, additional output would further invigorate domestic economies with increased demand for local commodities and labor.[37]

Community-level efforts to improve women farmers' status and livelihoods can become more effective if there are similar initiatives at the national scale. Policies governing assets, employment, and mobility can be altered to protect women's diverse needs and interests, including retention of joint property upon widowhood and freedom for sole caregivers to work in non-domestic employment or travel without male supervision in order to support their families.[38] Improved property or inheritance rights must go hand in hand with supporting measures to ensure and develop women's capacities to use their land or agricultural assets.[39] The Nicaraguan government recently undertook an interdepartmental effort to make the property legalization process easier for women to navigate through sensitization training for officials, and it launched campaigns to raise awareness about women's land rights among men and women.[40] In Mozambique, the government collaborated with civil society organizations to promote legal literacy among men and women by incorporating land legislation into literacy programs.[41]

Although 85 percent of countries made progress toward gender equity over the past seven years, according to the World Economic Forum's *Global Gender Gap Report 2011*, women farmers are still largely marginalized by development policies that are inattentive to their needs.[42] Current data are limited in scope and slow research efforts by not reflecting the wealth of knowledge and expertise that women are already using to, for example, mitigate global climate change. Food insecurity and climate change, along with associated trends such as land grabbing, large-scale biofuel production, and gendered migration and employment patterns, are also putting increasing pressure on women farmers to produce more with fewer resources.[43]

Developing a rights-based policy framework requires collaborative research, learning, and action within the international community for a global movement to empower women farmers with the resources, support, and recognition they need and deserve.

Foreign Investment in Agricultural Land Down from 2009 Peak

Cameron Scherer

Since 2000, an estimated 70.2 million hectares (ha) of agricultural land have been sold or leased to private and public investors.[1] This is a land mass roughly the size of the Democratic Republic of Congo and is 1.4 percent of the world's agricultural land.[2] The bulk of these acquisitions, which are called "land grabs" by some observers, took place between 2008 and 2010 (the most recent year for which data are available), peaking in 2009. Although data for 2010 indicate that the area acquired dropped considerably after the 2009 peak, the figure still remains well above pre-2005 levels.[3] (See Figure 1.)

Though definitions vary, "land deal" here refers to the large-scale purchase of agricultural land by public or private investors. In April 2012, the Land Matrix Project—a global network of some 45 research and civil society organizations—released the largest database to date on these types of land deals, gathering data from 1,006 deals covering 70.2 million ha around the world.[4]

The data remain in many ways lacking: with only a limited amount of information available, these statistics are admittedly conservative estimates. For some deals, little is known (for example, data on the date of contract are available for only 54.7 million ha of land acquisitions).[5] Furthermore, countries marked by open and transparent government may be overrepresented in the database, and the steep decline in deals following 2009 may reflect both reduced investment and waning media interest in tracking land grabs.

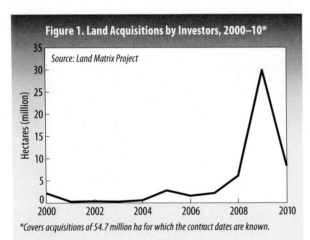

Figure 1. Land Acquisitions by Investors, 2000–10*

Source: Land Matrix Project

*Covers acquisitions of 54.7 million ha for which the contract dates are known.

Nevertheless, this database provides the first comprehensive look at the nature of these deals. It represents a significant expansion of a dataset released in January 2012 by GRAIN—a Barcelona-based nonprofit organization that supports community-controlled and biodiversity-based food systems—which contained information on 416 deals covering 35 million ha.[6]

Africa has seen the greatest share of land involved in these acquisitions, with 34.3 million ha sold or leased since 2000.[7] According to a report from the Land Matrix Project released at the same time as the database, which analyzed an additional 211 publicly reported deals, some 56.2 million ha have been sold or leased in Africa—4.8 percent of the continent's agricultural land.[8] East Africa accounts for the greatest in-

Cameron Scherer was a marketing and communications associate at Worldwatch Institute.

vestment, with 310 deals covering 16.8 million ha.[9] Increased investment in Africa's agricultural land reflects a decade-long trend of strengthening economic relationships between Africa and the rest of the world, with foreign direct investment to the continent growing 259 percent between 2000 and 2010.[10] (See Figure 2.)

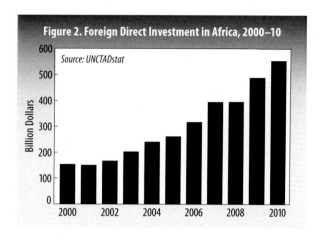

Figure 2. Foreign Direct Investment in Africa, 2000–10

Source: UNCTADstat

Asia and Latin America come in second and third for most heavily targeted regions, with 27.1 million and 6.6 million hectares of land deals, respectively.[11] Indonesia (at 9.5 million ha), the Philippines (5.2 million), Malaysia (4.8 million), and India (4.6 million) take four of the top six slots of individual target nations.[12] (See Table 1.)

Investor countries, in contrast, are spread more evenly around the globe. Still, patterns are discernible. Emerging economies are the largest bloc of investor countries. Of the 82 listed investor countries in the Land Matrix Project database, Brazil, India, and China account for 16.5 million ha—which is 23.5 percent of the total hectares sold or leased worldwide.[13] When the East Asian nations of Indonesia, Malaysia, and the Republic of Korea are included, this group of industrializing countries has been involved in 274 land deals covering 30.5 million ha.[14]

The United States and the United Kingdom account for a combined 6.4 million ha of land deals.[15] The oil-rich but arid Gulf states make up the final group of major land investors, with Saudi Arabia, the United Arab Emirates, and Qatar responsible for 4.6 million ha.[16]

Table 1. Top 10 Countries Where Land Deals Took Place, as of April 2012			
Country	Region	Hectares	Deals
Indonesia	Southeast Asia	9,527,760	24
Dem. Rep. of the Congo	Central Africa	8,051,870	10
Ethiopia	Eastern Africa	5,345,228	83
Philippines	Southeast Asia	5,182,021	45
Malaysia	Southeast Asia	4,819,483	20
India	South Asia	4,628,578	113
Sudan	Northern Africa	3,923,430	18
Brazil	South America	3,871,824	61
Madagascar	Eastern Africa	3,779,741	39
Zambia	Eastern Africa	2,273,413	9

Source: Land Matrix Project, Land Matrix Project Database, at www.landportal.info/landmatrix, updated April 2012.

In several cases—namely, South Africa, China, Brazil, and India—there is an overlap between investor and target countries. Yet most of the data paint one of two pictures. First, there is a new "South-South" regionalism, in which emerging economies invest in nearby, culturally affiliated countries.[17] Indeed, 32 percent of publicly recorded transactions occurred between countries within a given region.[18] This was particularly the case in Asia, where this category accounts for as much as 57 percent of land acquisitions.[19]

The other trend is one of wealthy (or increasingly wealthy) countries, many with little arable land, buying up land in low-income nations—especially those that have been particularly vulnerable to the financial and food crises of recent years. Overall, investor countries had a per capita gross domestic product five times that of target countries.[20] Moreover, in 2009 the 52 target-only countries (in other words, those with no domestic investors in foreign land) imported a net $856 million worth of food.[21] Because so few details of these transactions are made available to the public, these figures prompt observers to worry that in poor, weakly governed countries already dependent on a volatile world market for food imports, these land acquisitions are making it even more difficult for local people to meet their own food needs.

Of the 658 deals with information on individual investors, 442 deals (67 percent) were carried out by private companies, followed by public or state-owned investors in 172 cases (26 percent), investment funds in 32 deals (5 percent), and public-private partnerships in 12 deals (just under 2 percent).[22] Once again, regional discrepancies emerge: investors in North America, South America, and Europe are predominantly private companies, whereas in the Gulf States (excluding Saudi Arabia) and several Asian countries, public or state-owned actors are the driving factor.[23]

Data on the industry of the primary investor are available for 963 of the Land Matrix Project's documented deals; of these, 690 (72 percent) are in the agricultural sector.[24] (See Figure 3.) This number is significantly higher than the 52 percent reported in GRAIN's January database.[25] Just over one quarter of the acquired land is therefore used for nonagricultural purposes: some 11 percent of investors are in the forestry sector, and 8 percent are from the mining, industry, livestock, or tourism sectors.[26] (The source of the remaining 9 percent is unknown.)[27] The predominance of agricultural actors in land deals speaks to both the nature of these investments and the driving force behind them.

The food crisis of 2007–08 clearly helped spark the dramatic uptick in acquisitions in 2009, as investors rushed to capitalize on the rising prices of staple crops.[28] But food prices are not solely responsible for the recent trend. The worldwide demand for biofuels plays a key role as well. Although much of the information regarding the specific purpose of each investment remains shrouded in uncertainty, given that certain crops can serve as both food products and biofuels inputs, the data do make it clear that food-only crops account for only 26 percent of sold or leased land.[29]

As fuel consumption and oil prices continue to rise, the demand for biofuels will likely rise too, particularly given increasingly stringent emissions and renewable fuel targets implemented by governments around the world. The European

Union, for example, has set a target of 10 percent of its transport fuels being met by renewable energy by 2020.[30] Biofuels are expected to meet 80–90 percent of this demand.[31] Thus although 2010 saw a significant reduction in the number of reported land deals, the twin constraints on food (2011 saw a resurgence of high prices) and energy resources—and the level of media attention they are likely to draw to the issue—make it very possible that the next several years will see a spike in land acquisitions similar to that of 2009.[32]

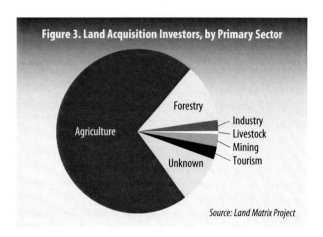

Figure 3. Land Acquisition Investors, by Primary Sector

Source: Land Matrix Project

Still, other factors are also at play. The Land Matrix Project reports that less than 30 percent of reported deals are believed to be in production, fueling rumors that many of these investments are purely speculative.[33] With the financial crisis of 2008 demonstrating the unreliability of global bond and equity markets, the steadily growing rates of return on agricultural land and land-based commodities are increasingly attractive.[34] Finally, the pressure to find new sources of livestock, timber, and water reservoirs to meet the needs of a growing population (and a worldwide middle class) is another driving force behind the global land deals.[35]

The implications of the recent surge in land acquisitions are still unclear. On the one hand, a 2010 World Bank report drew on statistical databases and satellite imagery to conclude that the world still contained over 445 million ha of unused land.[36] This suggests that acquisition of idle land could potentially serve to increase a nation's agricultural productivity.

On the other hand, as a result of ambiguous land rights and inaccessible legal institutions, these deals in many cases displace local farmers who already occupy and farm the land, although often without a formal legal document.[37] Furthermore, the capital-intensive agriculture that often results from these deals is another step away from the inclusive, sustainable agriculture needed to feed a growing population.[38] Absent regulations that go beyond voluntary guidelines, robust enforcement mechanisms, government transparency, and channels for civil society participation, further investments in land may benefit a group of increasingly wealthy investors at the expense of those living in the targeted countries.

Global Economy and Resources Trends

Control room of a moving grate incinerator for municipal solid waste, Germany

For additional global economy and resources trends, go to vitalsigns.worldwatch.org.

Wage Gap Widens as Wages Fail to Keep Pace with Productivity

Michael Renner

The economic crisis in 2008 was one of the harsher signs that economic globalization has gone hand in hand with increased volatility and turbulence. It caused the ranks of the unemployed to swell from 169 million in 2007 to 198.4 million in 2009, according to the International Labour Organization (ILO).[1] Although the number temporarily dipped to 193.1 million in 2011, a preliminary estimate for 2012 indicated it was back up to 197.3 million.[2] And the number of workers in vulnerable employment globally was estimated at close to 1.54 billion in 2012—about 55 percent of total employment worldwide—up from 1.39 billion in 2000.[3]

For the global workforce as a whole, the crisis has translated into a slowdown of wage growth, from an average of 3 percent in 2007 to 2.1 percent in 2010 and then to 1.2 per cent in 2011.[4]

Cumulatively, from 2000 to 2011 global real monthly average wages grew by just under one quarter.[5] (See Figure 1.) But global figures hide considerable regional differences. Wages almost doubled in Asia, whereas they increased by 18 percent in Africa and 15 percent in Latin America and the Caribbean.[6] Wages in the Eastern Europe and Central Asia region (which includes Russia) nearly tripled.[7] But this surge came on the heels of economic collapse after the fall of communism, which led wages to contract severely. In Russia, the subsequent growth only returned wages to what they had been at the beginning of the 1990s.[8] In the Middle East, the limited wage data available suggest stagnation during the last decade.[9] In industrial economies, wages increased by a comparatively tiny 5 percent, albeit from a much higher base than in other parts of the world.[10]

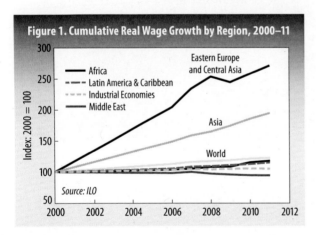

Figure 1. Cumulative Real Wage Growth by Region, 2000–11

Index: 2000 = 100

Africa
Latin America & Caribbean
Industrial Economies
Middle East

Eastern Europe and Central Asia

Asia

World

Source: ILO

Data collected by the U.S. Bureau of Labor Statistics (BLS) for the manufacturing sectors of 34 countries illustrate the tremendous wage differentials around the world. (See Figure 2.) Countries with the highest hourly compensation are primarily found in northern and western Europe; Norway had the highest reported compensation at $64.15 per hour in 2011.[11] Japan and the United States are in the middle of the field, while southern and eastern European countries, most of Asia, and Latin America all have lower compensation.[12] The Philippines has the lowest rate of the 34 countries, at $2.01.[13]

In all countries included in the BLS report (except Greece, where wages plum-

Michael Renner is a senior researcher at Worldwatch Institute.

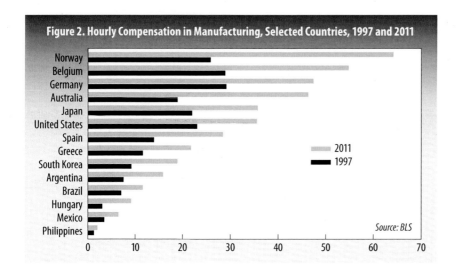

Figure 2. Hourly Compensation in Manufacturing, Selected Countries, 1997 and 2011

meted in the face of austerity policies), hourly compensation expressed in the respective local currencies rose in 2011.[14] But the numbers in Figure 2 are influenced by the exchange rates with the U.S. dollar.[15] Everywhere except Argentina, the local currency appreciated vis-à-vis the dollar, which has the effect of raising values expressed in dollars.[16]

The BLS data do not include China and India. BLS does separately offer estimates for these two countries ($1.36 per hour for China in 2008 and $1.17 for India in 2007), but due to data gaps and methodological issues these figures cannot be directly compared with the 34 countries the BLS reported on.[17] The estimate for India covers only the country's formal manufacturing sector, for example. People who work in the informal manufacturing sector account for some 80 percent of India's total manufacturing employment, but they earn substantially less than workers in the formal sector.[18]

Since the 1980s—long before the world economic crisis of 2008—wages in many countries stopped keeping pace with improvements in labor productivity.[19] Trade globalization, the expansion of financial markets, and declining trade union membership combined to erode the bargaining power of workers, and thus less of the wealth produced globally is going to labor compensation while a rising share is going to profits.[20] According to the ILO, average labor productivity in industrial countries increased more than twice as much as average wages did between 1999 and 2011.[21] (See Figure 3).

These trends are particularly pronounced in Germany. Real monthly wages remained flat during the past two decades, even as productivity grew by almost 23 percent.[22] This period of time saw a massive shift from full-time jobs to lower-pay part-time employment. The former fell by 4.9 million (16.6 percent) between 1991 and 2012, while part-time work expanded by 7.9 million jobs (167.7 percent).[23] In addition, the number of "Leiharbeiter" (literally, workers on loan who are sent to employers by private agencies) soared from 176,000 jobs in 1996 to almost 900,000 in 2011. Even though they are working de facto full-time, these individu-

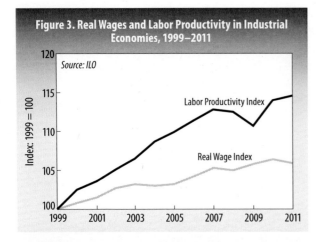

Figure 3. Real Wages and Labor Productivity in Industrial Economies, 1999–2011

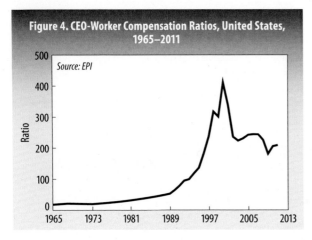

Figure 4. CEO-Worker Compensation Ratios, United States, 1965–2011

als are typically not well paid and receive no benefits.[24]

Average wage figures mask extremes of wage inequality. There is in fact a growing gap between top earners and the rest of the work-force, especially in terms of those in unskilled or low-skilled, temporary, precarious jobs.[25] The ILO finds that especially in English-speaking countries there has been a sharp increase in the salaries and compensation of top corporate executives.[26] In the United States, the top 1 percent of wage earners saw their annual earnings go up by 156 percent between 1979 and 2007.[27] For 90 percent of U.S. workers, in contrast, wages advanced by a much smaller 17 percent during the same period.[28]

In particular, the wage discrepancy between chief executive officers (CEOs) and average workers has reached all-time highs in recent years. From 1978 to 2011, CEO compensation (including salaries, bonuses, long-term incentive pay, and stock options) at the 350 largest U.S. companies increased more than 725 percent—compared with just 5.7 percent in average worker compensation.[29] (See Figure 4.) By a measure that includes the value of stock options granted, the CEO-to-worker compensation ratio rose from 18 to 1 in 1965 to a peak of 411 to 1 in 2000, and it was 209 to 1 in 2011.[30]

The gap between wages and labor productivity and the rising inequality of wages are developments that raise fundamental questions of fairness in the economy. The extremely unequal distribution of income and wealth that has emerged worldwide has profound consequences, determining who has an effective voice in matters of economics and politics—and thus how countries address the fundamental challenges before them.

But wage inequality also has deleterious impacts on economic stability. In the United States, for example, in response to stagnating incomes, many people reduced their savings and increased consumer debt. This included leveraging exaggerated real estate values. When the resulting bubble burst, it was a major contributor to the global economic crisis.[31]

Evidence from Europe also suggests potentially dangerous impacts on economic stability. Among the 12 members of the European Union in the period 1999–2010, Germany was the only country where average real compensation per employee declined.[32] This policy of "wage dumping" caused domestic demand to be weak, but it allowed Germany to boost its exports to other European countries.[33] As a result, these countries felt rising pressure to lower their own labor costs.[34] Yet it is impossible for all countries to adopt the same policy. Competitive wage cuts pursued simultaneously by many countries could trigger a race to the bottom.[35]

For reasons of fairness and stability, today's growing wage gaps cannot continue indefinitely. There are no easy solutions in an ever-more-interconnected global economy. But a return to stronger worker representation in industrial economies and growing workers' rights in developing economies are essential ingredients of change. The ILO points to the importance of collective bargaining and minimum wages to help achieve a more balanced and equitable recovery from economic crisis.[36] And the organization finds that in the countries where collective bargaining agreements cover more than 30 percent of all employees, productivity and wage trends are more in tune with each other than in other countries. Collective bargaining and minimum wages also tend to reduce the share of workers earning low wages.[37]

Metals Production Recovers

Gary Gardner

Global production of key metals surged 14.3 percent in 2010 (the latest year with data) to a record 1.48 billion tons, in a robust recovery from the sharp decline spurred by the 2009 global recession.[1] (See Figure 1.) The increase marks a return to the steep rise in metals production of the past decade, driven in part by the rapid economic expansion of newly prosperous developing countries such as China, India, and Brazil.

The metals covered in Figure 1 are common ones of great economic importance: aluminum, arsenic, cadmium, chromium, copper, gold, lead, mercury, nick-

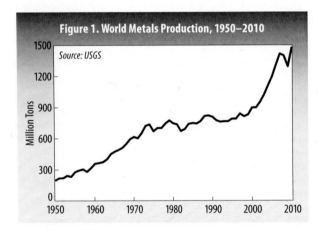

Figure 1. World Metals Production, 1950–2010

Source: USGS

el, and steel.[2] But dozens of other metals also show strong growth as more countries industrialize and as increasingly complex products use metals not heard about in daily conversation, from hafnium used in nuclear reactors to rhenium, a key metal in jet engines.

The surge in metals production continues the trend of the past half-century or more. Global extraction growth rates for copper, zinc, nickel, tin, and platinum averaged about 3.4 percent annually over the past few decades, which implies a doubling of extracted material volumes every 20 years.[3] Indeed, humanity's intensive use of materials was in a league of its own in the twentieth century: by one estimate, 97.5 percent of all copper produced worldwide in the last 1,000 years has been produced since 1900.[4]

The surge of the past decade is spurred by robust economic output in advanced developing countries. Steel, a metal strongly associated with infrastructure development and with industries like the automobile, is emblematic of the trend. Whereas steel production declined sharply in advanced industrial regions during the 2007–09 recession, it continued its longtime increase in Asia, driven in particular by ongoing strong growth in China.[5] In 2001, China's steel production was about 50 percent greater than that of the world's second largest producer, Japan.[6] Then in the past decade, China's output quadrupled, and by 2010 it was 5.7 times greater than that of Japan.[7] (See Figure 2.) Indeed, China ranks in the top five producers globally for 8 of the 10 metals covered in Figure 1.[8]

Industrial regions like North America and the European Union traditionally have high rates of metals consumption per person, far beyond those in developing

Gary Gardner is director of grants administration at the Resources Legacy Fund in California.

nations.[9] A 2010 U.S. Geological Survey study of aluminum found that countries whose gross domestic product (GDP) per person is less than $5,000 consume less than 5 kilos of aluminum per person; those whose GDP per person falls between $5,000 and $15,000 consume 5–10 kilos of aluminum per person; but a per capita income of greater than $25,000 typically implies consumption of between 15 and 35 kilos.[10] Not surprisingly, wealthier countries have greater levels of stocks in use—in buildings, vehicles, and myriad other economic outputs—than in developing countries, often by a factor of 3 to 10.[11] (See Table 1.)

The gap may be narrowing, however, as some developing economies mature. In the recent recession, Asian per capita consumption of steel continued upward, for example, while U.S. consumption dipped sharply.[12] (See Figure 3.) Wealthy-nation consumption per person could well recover, but robust growth in developing countries that have extensive infrastructure and other needs means that even at a per capita level of steel consumption—long an area of industrial-country dominance—advanced developing countries now lead the world.

The trend holds for aluminum as well. Most of the nearly 2.7-fold increase in world aluminum consumption between 2006 and 2025 is expected to come from developing countries.[13] Brazil, Russia, India, and China, which accounted for 26 percent of world consumption in 2006, are expected to account for 45 percent in 2025.[14] Meanwhile, the global share of aluminum consumption represented by the United States, Japan, Germany, and France is expected to fall from 36 percent in 2006 to 15 percent in 2025.[15]

For some metals, the combination of long-standing wealthy-country demand and surging demand in populous developing countries could begin to approach the limits of available supply. Researchers have estimated, for example, that global in-ground stocks of copper total about 1,600 billion kilos.[16] Comparing this to copper consumption in North America—about

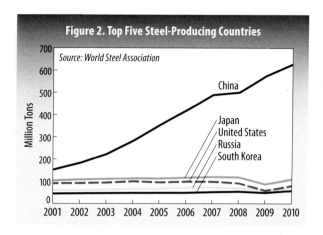

Figure 2. Top Five Steel-Producing Countries

Source: World Steel Association

China
Japan
United States
Russia
South Korea

Table 1. In-use Stocks of Five Metals, per Person, Industrial and Developing Countries

Metal	Industrial Countries	Developing Countries
	(kilograms)	
Aluminum	350–500	35
Copper	140–300	30–40
Iron	7,000–14,000	2,000
Lead	20–150	1–4
Zinc	80–200	20–40

Source: T. E. Graedel, Metal Stocks in Society: A Scientific Synthesis (Paris: International Resource Panel, UNEP, 2010).

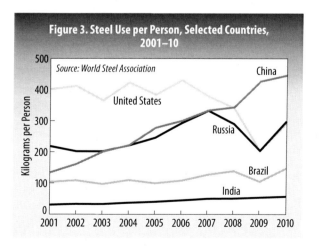

Figure 3. Steel Use per Person, Selected Countries, 2001–10

Source: World Steel Association

China
United States
Russia
Brazil
India

170 kilos per person—they conclude that a global population of 10 billion people (the possible world total in the early second half of this century) could require 1,700 billion kilos, or more than the estimated total global supply.[17] Similarly for platinum, which is used in fuel cells that might propel cars of the future: powering the world's half-billion or so cars using fuel cells would deplete the world's platinum supply in about 15 years, even if half the platinum in those cells were recycled.[18] This research suggests that for many metals it is not too soon to set policies to conserve supplies.

Yet progress in resource conservation is slow. The U.N. Environment Programme's International Resource Panel noted in May 2011 that for only 18 of 60 metals is the share of discarded metal that is recycled—known as the end-of-life recycling rate—above 50 percent.[19] (See Table 2.) High rates tend to be associated with materials that are used in large amounts in easily recoverable applications, such as car manufacturing.[20] Metals with lower rates of recovery are often used in small quantities in complex products, such as electronics.[21] Recycling rates are especially low for specialty metals that are used in emerging technologies.[22]

Table 2. End-of-Life Recycling Rates for 60 Metals

Recycling Rate	Number of Metals	Metals
Greater than 50 percent	18	Aluminum, cobalt, chromium, copper, gold, iron, lead, manganese, niobium, nickel, palladium, platinum, rhenium, rhodium, silver, tin, titanium, zinc
25 to 50 percent	3	Magnesium, molybdenum, iridium
10 to 25 percent	3	Ruthenium, cadmium, tungsten
1 to 10 percent	2	Antimony, mercury
Less than 1 percent	34	Lithium, beryllium, boron, scandium, vanadium, gallium, germanium, arsenic, selenium, strontium, yttrium, zirconium, indium, tellurium, barium, hafnium, tantalum, thallium, bismuth, lanthanum, cerium, praseodium, neodymium, samarium, europium, gadolinium, terbium, dysprosium, holmium, erbium, thulium, ytterbium, lutetium, osmium

Source: Thomas Graedel et al., Recycling Rates of Metals: A Status Report, A Report of the Working Group on Global Metal Flows *(Paris: International Resource Panel, UNEP, 2011).*

Indeed, economies based on one-way flows of materials hold extensive metals stocks in landfills. The U.S. Geological Survey (USGS) noted in 2005 that aluminum in dumps in the United States was equal to about 43 percent of the aluminum in use in the U.S. economy.[23] For copper, stocks in dumps were about 13 percent of stocks in use in the country.[24] And for steel, landfill stocks amounted to about 20 percent of all steel in use in the United States.[25] In fact, the USGS estimates that U.S. landfills hold enough steel to build 11,000 Golden Gate Bridges.[26] Recovery of these landfill-based resources is not yet economically feasible, but it might be one day.

Future rates of recycling will be affected in part by the patterns of metals use in developing countries, which are different than in industrial ones. For example, developing countries tend to use a larger share of aluminum in electrical systems than richer countries and a smaller share in transportation.[27] Moreover, aluminum may reside in electrical systems for decades but for perhaps only 10 years in a car. So developing countries will generate less aluminum scrap for recycling in the early decades of development than in the mature stages of their economic growth.[28] Thus, even policies that promote recycling of metals will need to take into account the life-cycle patterns of metals use in a particular economy.

Materials researchers increasingly outline policy options that can help economies get more service out of every ton of metal. Above all, they cite the need to create a circular economy by reusing and recycling materials and by remanufacturing products to the extent possible. Germany, Japan, and China are leaders in making a circular economy a priority. In Japan, for example, a steady progression of waste reduction laws over the past 20 years has helped to create just such an economy.[29]

Municipal Solid Waste Growing

Gary Gardner

Some 1.3 billion tons of municipal solid waste (MSW) are generated globally each year, a volume that is increasing rapidly as urbanization, mass consumption, and throw-away lifestyles become more prevalent worldwide.[1] The volume of MSW generated globally is projected to double by 2025 as two drivers of garbage generation—prosperity and urbanization—continue to advance, particularly in developing countries.[2] The trend poses serious environmental and health challenges to cities worldwide.[3] To the extent that MSW is not treated as a resource—and in most countries it is not—it stands as an indicator of economic unsustainability.

As used here, MSW consists of organic material, paper, plastic, glass, metals, and other refuse collected by municipal authorities, largely from homes, offices, institutions, and commercial establishments.[4] MSW is a subset of the larger universe of waste. It typically does not include waste collected outside of formal municipal programs. Nor does it include the sewage, industrial waste, or construction and demolition waste generated by cities.[5] And of course MSW does not include rural wastes. The U.N. Environment Programme (UNEP) estimates that MSW and industrial wastes combined amount to between 3.4 billion and 4 billion tons a year—roughly three times greater than the flow of MSW, the focus here.[6] MSW is measured at collection, so data on it often include collected material that is later diverted for recycling.

MSW tends to be generated in much higher quantities in wealthier regions of the world. Members of the Organisation for Economic Co-operation and Development (OECD), a group of 34 industrial nations, lead the world in MSW generation, at nearly 1.6 million tons per day.[7]

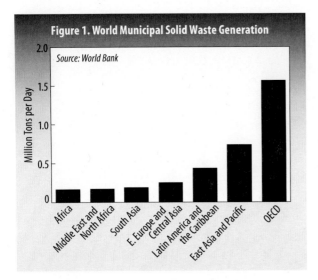

Figure 1. World Municipal Solid Waste Generation

Source: World Bank

Million Tons per Day

Africa; Middle East and North Africa; South Asia; E. Europe and Central Asia; Latin America and the Caribbean; East Asia and Pacific; OECD

(See Figure 1.) By contrast, sub-Saharan Africa produces less than one eighth as much, some 200 million tons per day.[8]

The list of top 10 MSW-generating countries includes 4 developing nations, in part because of the size of their urban populations and in part because their city dwellers are prospering and adopting high-consumption lifestyles.[9] (See Table 1.) While the United States leads the world in MSW output at some 621,000 tons per day, China is a relatively close second, at some 521,000 tons. Even among the top

Gary Gardner is director of grants administration at the Resources Legacy Fund in California.

10, however, there is a wide range of output: the United States generates nearly seven times more urban refuse than France, in tenth position, does.

On a per person basis, some 40 of the top 50 countries in waste generation are high-income countries.[10] (See Figure 2.) (This ranking excludes 10 nations in which the ratio of tourists to residents is high. Large numbers of waste-generating visitors boost a city's waste generation rates to uncommonly high levels.) OECD nations generate the greatest quantities of garbage, more than 2 kilograms per person per day.[11] In South Asia, the rate is less than one quarter as high, under half a kilo per person.[12] People in the Middle East and North Africa, in Latin America and the Caribbean, and in Eastern Europe and Central Asia generate just over 1 kilo of MSW per person each day.[13]

Urbanization and income levels also tend to determine the type of waste generated. The share of inorganic materials in the waste stream, including plastics, paper, and aluminum, tends to increase as people grow wealthier and move to cities. Waste flows in rural areas, in contrast, are characterized by a high share of organic matter, ranging from 40 to 85 percent.[14] Similarly, organic waste accounts for more than 60 percent of MSW in low-income countries, but only a quarter of the waste stream in high-income countries.[15] (See Figure 3.)

Roughly a quarter of the world's garbage is diverted to recycling, composting, or digestion—waste management options that are environmentally superior to landfills and incinerators. Recycling rates vary widely by country.[16] In the United States, the recycled share of MSW grew from less than 10 percent in 1980 to 34 percent in 2010, and similar increases have been seen in other countries, especially industrial ones.[17] For countries with data, official statistics suggest that recycling is largely an industrial-country practice.[18] (See Table 2.) In low-income countries, however, recovery and recycling often occur in the informal sector and may not be recorded in formal statistics. In China, for example, some 20 percent of discarded products and materials is estimated to be recovered for recycling by poor "waste pickers."[19] And UNEP's Green Economy report suggests that recycling rates in the

Table 1. Top 10 MSW-Generating Countries

Country	Total MSW Generation	MSW Generation per person
	(tons/day)	(kilos/person/day)
United States	621,232	2.58
China	520,548	1.02
Brazil	149,096	1.03
Japan	144,466	1.71
Germany	127,816	2.11
India	109,589	0.34
Russia	100,027	0.93
Mexico	99,014	1.24
United Kingdom	97,342	1.79
France	90,493	1.92

Source: Daniel Hoornweg and Perinaz Bhada-Tata, What a Waste: A Global Review of Solid Waste Management *(Washington, DC: World Bank, 2012); U.S. EPA, "Municipal Solid Waste Generation, Recycling, and Disposal in the United States: Facts and Figures for 2010."*

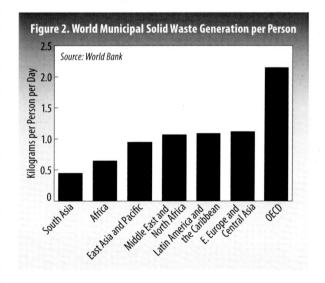

Figure 2. World Municipal Solid Waste Generation per Person

Source: World Bank

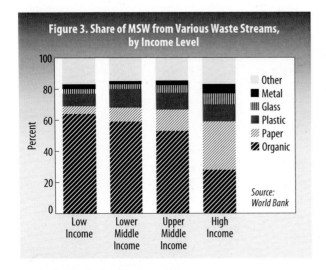

Figure 3. Share of MSW from Various Waste Streams, by Income Level

Source: World Bank

Legend: Other, Metal, Glass, Plastic, Paper, Organic

Table 2. Top 15 Recyclers of MSW, 2010

Country	Recycled Share
	(percent)
South Korea	49
Singapore	47
Hong Kong	45
Ireland	34
Norway	34
Sweden	34
Switzerland	34
United States	34
Belgium	31
Marshall Islands	31
Australia	30
Canada	27
Austria	27
Denmark	26
Netherlands	25

Source: Daniel Hoornweg and Perinaz Bhada-Tata, What a Waste: A Global Review of Solid Waste Management (Washington, DC: World Bank, 2012); U.S. EPA, "Municipal Solid Waste Generation, Recycling, and Disposal in the United States: Facts and Figures for 2010."

informal sector in developing countries could be in the 20–50 percent range, although it cautions that such estimates need validation.[20]

The growing interest in MSW recovery is driven by a maturation of regulations and of markets for post-consumer materials. The global market for scrap metal and paper is at least $30 billion per year, according to the World Bank.[21] UNEP estimates the market for waste management, from collection through recycling, to be more like $400 billion worldwide. Figures like these are bound to get the attention of investors and the waste industry.[22] Yet UNEP also estimates that to "green" the waste sector would require, among other things, a 3.5-fold increase in MSW recycling at the global level, including nearly complete recovery of all organic material through composting or conversion to energy.[23]

Recycling saves virgin materials, reduces the environmental impacts of logging and mining, and saves energy. The U.S. Environmental Protection Agency estimates that recycling 8 million tons of metals in the United States has eliminated more than 26 million tons of greenhouse gases—the equivalent of removing more than 5 million cars from the road for a year.[24] One ton of aluminum produced from recycled aluminum uses 95 percent less energy than one ton made from virgin ore, in addition to saving more than 15 cubic meters of water and avoiding emissions of 2 tons of carbon dioxide and 11 kilos of sulfur dioxide.[25] And each ton of recycled paper saves 17 trees and the energy equivalent of 165 gallons of gasoline compared with paper made from trees, in addition to requiring only half the water.[26]

The gold standard for MSW will be to integrate it into a materials management approach known as a "circular economy," which involves a series of policies to reduce the use of some materials and to reclaim or recycle most of the rest. Japan has made the circular economy a national priority since the early 1990s through passage of a steady progression of waste reduction laws, and the country has achieved notable successes.[27] Resource productivity (tons of material used per yen of gross domestic product) is on track to more than double by 2015 over 1990 levels, the recycling rate is projected to roughly double over the same period, and total material sent to landfills will likely decrease to about one fifth the 1990 level by 2015.[28]

Losses from Natural Disasters Reach New Peak in 2011

Petra Löw

During 2011, a total of 820 natural catastrophes were documented, a decrease of 15 percent from the 970 events registered in 2010.[1] But the 2011 figure is in line with the average of 790 events during 2001–10 and is considerably above the average of 630 events during 1981–2010.[2] (See Figure 1.)

The breakdown of loss-relevant events among the main hazards—geophysical, meteorological, hydrological, and climatological events—is more or less in line with the average over the past 30 years.[3] In 2011, some 91 percent were weather-related—37 percent each were storms and floods and 17 percent were climatological events like heat waves, cold waves, wildfires, and droughts—while 9 percent were geophysical events, including earthquakes, tsunamis, and volcanic eruptions.[4]

The share of events by continent is also in line with the long-term average. Most natural catastrophes occurred in the Americas (290) and Asia (240), while in Europe there were 150 such events, 80 in Africa, and 60 in Australia.[5]

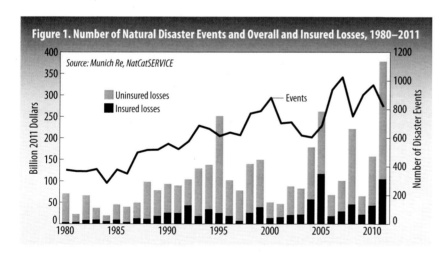

Figure 1. Number of Natural Disaster Events and Overall and Insured Losses, 1980–2011

Source: Munich Re, NatCatSERVICE

In 2011, an estimated 27,000 people died in natural catastrophes, which was 63 percent below the long-term average of 73,000 fatalities per year (1980–2010).[6] The 2011 death toll is only 9 percent of the toll of the deadliest year in this period, 2010, when there were 296,000 deaths.[7]

In 2011, some 62 percent of the fatalities were caused by geophysical events, almost all of them (15,840 deaths) accounted for by the earthquake and tsunami in Japan on March 11.[8] Another 11 percent of the fatalities were due to storms, 25

Petra Löw is a geographer and works as a consultant at Munich Reinsurance Company with a focus on natural catastrophe losses.

percent were caused by floods, and 2 percent were due to climatological events.[9] In total, 38 percent of all victims lost their lives in weather-related disasters.[10] (The number of fatalities in 2011 does not take into account the catastrophic drought and subsequent famine on the Horn of Africa, one of the year's worst humanitarian catastrophes, which is discussed later.)[11]

The year's overall losses climbed to $380 billion.[12] This is the highest figure ever recorded in Munich Re's database of natural catastrophes, and it far surpasses the previous record of $220 billion set in 2005.[13] Insured losses also reached a record high of $105 billion.[14] Some 61 percent of the overall losses and 47 percent of the insured losses were caused by geophysical events.[15] This is significantly higher than the long-term loss pattern since 1980, where on average only 22 percent of overall losses and 10 percent of insured losses were due to geophysical events.[16]

Weather-related events accounted for 39 and 53 percent, respectively, of overall and insured losses in 2011, compared with the long-term averages of 78 and 90 percent.[17] Nonetheless, 2011 was the second costliest year for weather-related disasters since 1980, after taking inflation into account.[18]

Table 1 lists the 10 most expensive natural catastrophes in terms of overall losses in 2011 as registered in the Munich Re NatCatSERVICE database.[19]

The Americas, with 35 percent (290 events), and Asia, with 29 percent (240 events), were the regions with the highest number of natural catastrophes in 2011, followed by Europe with 19 percent (150 events), Africa with 10 percent (80 events), and the greater region of Australia/Oceania with 7 percent (60 events).[20] (See Figure 2.)

The distribution of the world's overall losses of $380 billion by region shows

Table 1. Costliest Natural Catastrophes in 2011

Date	Event	Region	Overall Losses	Insured Losses	Fatalities
			(million dollars)		
11 March	Earthquake, tsunami	Japan	210,000	35,000–40,000	15,840
Aug–Nov	Floods	Thailand	40,000	10,000	813
22 Feb	Earthquake	New Zealand	16,000	13,000	181
22–28 April	Tornado outbreak	United States	15,000	7,300	350
20–27 May	Tornado outbreak	United States	14,000	6,900	178
2011	Drought	United States	8,000	2,400	–
22 Aug–2 Sept	Hurricane Irene	Caribbean, United States	7,400	5,600	55
Dec 2010–Jan 2011	Floods	Australia	7,300	2,405	35
18 April–23 May	Floods (Mississippi)	United States	4,600	500	9
3–5 April	Severe storm, thunderstorms	United States	3,500	2,000	9

Source: Munich Re, NatCatSERVICE database, 2012.

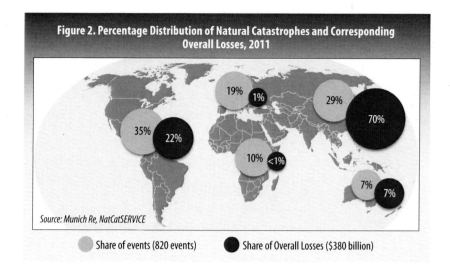

Figure 2. Percentage Distribution of Natural Catastrophes and Corresponding Overall Losses, 2011

19% 1%

29%

70%

35% 22%

10% <1%

7% 7%

Source: Munich Re, NatCatSERVICE

Share of events (820 events)   Share of Overall Losses ($380 billion)

that 70 percent of the losses occurred in Asia and nearly 22 percent in the Americas, mainly in North America and the Caribbean, followed by Australia/Oceania with 7 percent, Europe with 1 percent, and Africa with less than 1 percent.[21]

In the United States, an extremely active spring thunderstorm season caused damage on an unprecedented scale across the country. Numerous tornado outbreaks devastated entire cities and caused a record of $47 billion in overall losses, of which $26 billion were insured losses.[22] With 551 fatalities, the 2011 tornado season was the deadliest in the United States in more than 85 years.[23]

At the same time, the hurricane season in the Atlantic Ocean was the third-strongest since recordkeeping began, with 19 named storms.[24] Only 3 of these storms made landfall in the United States. But with overall losses of $7.4 billion and insured losses of $5.6 billion, Hurricane Irene alone stands amongst the 10 costliest events in 2011.[25] Over the past 10 years, annual hurricane losses have been close to $17 billion.[26] In May and June 2011, the worst floods in decades occurred along the Mississippi and Missouri Rivers and caused more than $5 billion in overall losses.[27]

In South America, 86 percent of all 2011 natural catastrophes were weather-related.[28] Severe flash floods and landslides affected Brazil in January and then Colombia between September and December. More than 1,400 people lost their lives.[29]

In Europe, 95 percent of the 150 disasters in 2011 were weather-related.[30] The overall loss of $2.5 billion is one of the lowest annual figures since 1980.[31] Torrential rains in northern Italy, southern France, and northeastern Spain triggered numerous landslides, killed 14 people, and caused $2.1 billion worth of damages, the single costliest event in Europe that year.[32]

The influence of the La Niña weather phenomenon from January to May and from August to December was a major cause of many of the extreme weather events in 2011.[33] One dramatic effect was the severe drought in the Horn of Africa from October 2010 until September 2011, causing widespread famine and large-scale migratory movements, particularly in Somalia and Kenya.[34] Around

80 percent of the livestock of the nomadic population in Somalia died, and some 13 million people required humanitarian aid.[35] An estimated 50,000 people lost their lives.[36]

Asia suffered overall losses of $265 billion in 2011 and saw its share of the worldwide total climb to 70 percent from the average of 38 percent for 1980–2010.[37] Two major natural disasters made the difference. On March 11, 2011, the most intensive earthquake ever recorded in Japan shook the northeast of the country.[38] The tsunami triggered by the quake not only devastated several hundred kilometers of coastline but also led to the nuclear disaster at Fukushima. The direct overall losses (excluding the nuclear accident) were estimated at $210 billion, of which $35–40 billion was insured.[39]

The second major event was the devastating flooding that hit Thailand from August to November.[40] In the north, 3.5 times the normal amount of rainfall in March led to numerous flash floods in the mountains.[41] Then the summer monsoon brought precipitation that was well above average over several months, probably influenced by a very intensive La Niña situation.[42] With $40 billion in overall losses and about $10 billion in insured losses, this was the costliest flood event to date.[43]

From December 2010 to January 2011, Australia was also severely affected by widespread floods, especially in the state of Queensland. The extreme precipitation there can be linked to La Niña patterns as well.[44] A loss of $7.3 billion makes this the costliest flood ever in Australia.[45] Shortly after the floods, Tropical Cyclone Yasi also hit Queensland and generated overall losses of $2.5 billion.[46] New Zealand suffered a series of earthquakes in 2011.[47] One in February caused overall losses of $16 billion, of which approximately 80 percent were insured, and a quake in June caused losses of up to $2 billion.[48]

Some 14 percent of all loss-relevant events in 1980–2011 were related to geophysical events, as were 26 percent of the overall losses and 12 percent of the insured losses.[49] (See Figure 3.) With 39 percent of the death toll from all disasters, geophysical events are the deadliest of hazards.[50] The last decade in particular was dominated by a series of devastating earthquakes and tsunamis with an enormous human impact—in 2010 in Haiti (222,570 deaths); in 2004 in Southeast Asia (220,000 deaths); in 2005 in Pakistan (88,000 deaths); and in 2008 in China (84,000 deaths).[51]

In 1980–2011, some 86 percent of all loss-relevant events were weather-related, mainly storms and floods.[52] The costliest weather catastrophes are tropical cyclones, floods, winter

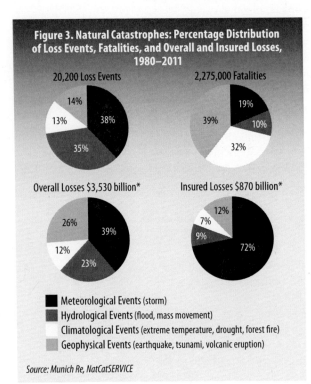

Figure 3. Natural Catastrophes: Percentage Distribution of Loss Events, Fatalities, and Overall and Insured Losses, 1980–2011

20,200 Loss Events

2,275,000 Fatalities

Overall Losses $3,530 billion*

Insured Losses $870 billion*

■ Meteorological Events (storm)
■ Hydrological Events (flood, mass movement)
□ Climatological Events (extreme temperature, drought, forest fire)
▨ Geophysical Events (earthquake, tsunami, volcanic eruption)

Source: Munich Re, NatCatSERVICE

storms, and thunderstorms.[53] Hurricane Katrina, which caused overall losses in 2005 of $125 billion and insured losses of $62.2 billion, is the most expensive weather catastrophe ever.[54] It is followed by the 2011 Thailand floods and by Hurricane Ike (in 2008). The latter had $38.3 billion in overall and $18.5 billion in insured losses.[55]

The deadliest weather catastrophes are droughts with consequent famines, especially in Africa.[56] The second deadliest weather events are tropical cyclones with storm surges, like Cyclone Nargis in Myanmar (in 2008), with 140,000 fatalities, and the 1991 cyclone and storm surge in Bangladesh in which 139,000 people lost their lives.[57]

The Looming Threat of Water Scarcity

Supriya Kumar

Some 1.2 billion people—almost one fifth of the world—live in areas of physical water scarcity, while another 1.6 billion face what can be called economic water shortage.[1] The situation is only expected to worsen as population growth, climate change, investment and management shortfalls, and inefficient use of existing resources restrict the amount of water available to people. It is estimated that by 2025, fully 1.8 billion people will live in countries or regions with absolute water scarcity, with almost half of the world living in conditions of water stress.[2]

Water scarcity has several definitions. Physical scarcity occurs when there is not enough water to meet demand; its symptoms include severe environmental degradation, declining groundwater, and unequal water distribution.[3] Economic water scarcity occurs when there is a lack of investment and proper management to meet the demands of people who do not have the financial means to use existing water sources; the symptoms in this case normally include poor infrastructure.[4] Large parts of Africa suffer from economic water scarcity. (See Figure 1.)

To measure water scarcity, hydrologists compare the size of a population with the amount of available water. According to the United Nations, an area is said to be experiencing water stress when annual water supplies fall below 1,700 cubic meters per person.[5] A region is said to face water scarcity when annual supplies per person fall below 1,000 cubic meters, and absolute water scarcity is when annual supplies per person drop below 500 cubic meters.[6]

Nearly all Arab countries are considered water-scarce, with consumption of water significantly exceeding total renewable supplies.[7] Twelve Arab countries have less than 500 cubic meters of renewable water resources available per person annually.[8] About 66 percent of Africa is arid or semiarid, and more than 300 million people in sub-Saharan Africa live on less than 1,000 cubic meters of water resources each per year.[9] (See Table 1.)

Although the Asia Pacific region is home to almost 60 percent of the world's population, it only has 36 percent of global water resources.[10] In 2009, the region had 2,970 cubic meters of water resources per person.[11] Although this is not a sign of water scarcity, it is still less than half of the world's average of 6,236 annual cubic meters.[12] Parts of northern China, India, and Pakistan suffer from both physical and economic scarcity. In comparison, the average amount of water available per person in Latin America is about 7,200 cubic meters, although it is only 2,466 cubic meters in the Caribbean.[13]

North America and Europe, in contrast, are well endowed with renewable water resources. Canada and the United States have about 85,310 and 9,888 annual cubic meters of water resources per person, respectively, while Europe has almost

Supriya Kumar is the communications manager at Worldwatch Institute.

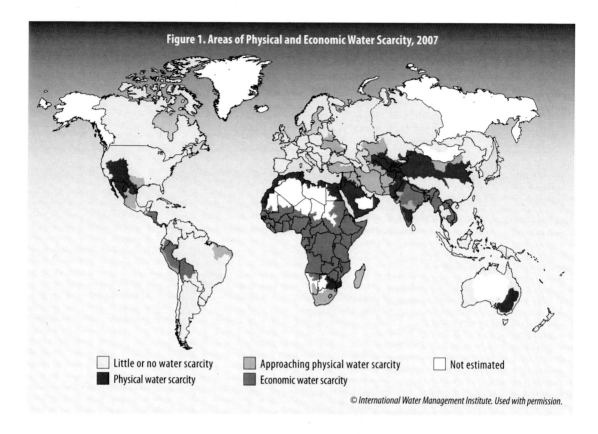

Figure 1. Areas of Physical and Economic Water Scarcity, 2007

☐ Little or no water scarcity   ☐ Approaching physical water scarcity   ☐ Not estimated
■ Physical water scarcity   ☐ Economic water scarcity

© International Water Management Institute. Used with permission.

4,741 cubic meters.[14] People in these regions also consume a considerable amount of "virtual water"—water that is used in the production of goods, especially agricultural products such as grain, which can then be traded. According to UN Water, each person in North America and Europe (excluding former Soviet Union countries) consumes at least 3 cubic meters per day of virtual water in imported food, compared with 1.4 cubic meters per day in Asia and 1.1 cubic meters per day in Africa.[15]

Water resources face many pressures, including population growth, increased urbanization and overconsumption, lack of proper management, and the looming threat of climate change. According to the U.N. Food and Agriculture Organization (FAO) and UN Water, global water use has been growing at more than twice the rate of population increase in the last century.[16] World population is predicted to

Table 1. Water Availability by Region, 2012	
Region	**Average Water Availability**
	(cubic meters per person)
Arab Countries	500
Sub-Saharan Africa	1,000
Caribbean	2,466
Asia Pacific	2,970
Europe	4,741
Latin America	7,200
North America (includes Mexico)	13,401

Source: FAO, AQUASTAT, at www.fao.org/nr/water/, viewed 1 March 2013; WWAP, World Water Development Report, Vol. 1: Managing Water under Uncertainty and Risk (Paris: UNESCO, 2012).

grow from 7 billion to 9.1 billion by 2050, putting a strain on water resources to meet increased food, energy, and industrial demands.[17]

At the global level, 70 percent of water withdrawals are for the agricultural sector, 11 percent are to meet municipal demands, and 19 percent are for industrial needs.[18] These numbers, however, are distorted by the few countries that have very high water withdrawals, such as China, India, and the United States. A breakdown of water withdrawal by sector and region is shown in Figure 2.

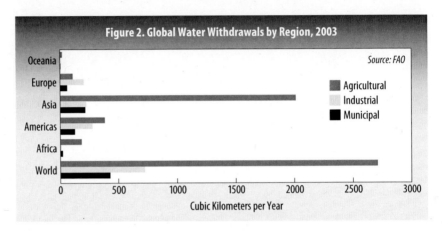

Figure 2. Global Water Withdrawals by Region, 2003

Agriculture is one the most water-intensive sectors, currently accounting for more than 90 percent of consumptive use.[19] Agricultural water withdrawal accounts for 44 percent of total water withdrawal among members of the Organisation for Economic Co-operation and Development (OECD), but this rises to more than 60 percent within the eight OECD countries that rely heavily on irrigated agriculture.[20] In the four transitional economies of Brazil, Russia, India, and China, agriculture accounts for 74 percent of water withdrawals, but this ranges from 20 percent in the Russia to 87 percent in India.[21]

Water use in agriculture is often inefficient, which has led to the overexploitation of groundwater resources as well as the depletion of the natural flow of major rivers, such as the Ganges in India and the Yellow River in China. Around 54 percent of the total area available for irrigation is irrigated with surface water, 5 percent with groundwater, and 41 percent with a combination of both sources.[22] But when both sources are used together, less than 15 percent of it is surface water, which has led to a doubling of the global depletion of groundwater resources in the last 50 years.[23] The Ganges, Indus, and Yellow River basins in Asia have already reached high levels of water crowding and suffer from sever water shortage due to overuse.[24]

Policymakers must introduce a variety of measures to address global water scarcity. One important initiative is to support small-scale farmers. Much of the public investment in agricultural water management has focused on large-scale irrigation systems. But supporting smallholder farmers, who in general operate without large infrastructure such as dams, canals, and distribution devices, can decrease the amount of water used in the agricultural sector. This support must be accompanied by smart subsidies. In India, for example, many farmers who receive free electricity all day are experiencing groundwater depletion due to overpumping.[25] To address this issue, policymakers in the state of Gujarat reduced the amount of time that farmers could pump water to eight hours on a pre-announced schedule that meets peak demand but also reduces total water usage.[26]

Farmers can also use water more efficiently by taking a number of steps, including growing a diverse array of crops suited to local conditions, especially in drought-prone regions; practicing agroforestry to build strong root systems and reduce soil erosion; maintaining healthy soils, either by applying organic fertilizer or growing cover crops to retain soil moisture; and adopting irrigation systems like "drip" lines that deliver water directly to plants' roots. Rice farmers, for example, can adopt the System of Rice Intensification (SRI), which not only increases crop yields but uses 20–50 percent less water than conventional rice production.[27] SRI is an innovative method of increasing the productivity of irrigated rice with very simple adjustments to traditional techniques. It involves transplanting younger seedlings into a field with wider spacing in a square pattern, irrigating to keep the roots moist and aerated instead of flooding fields, and increasing organic matter in the soil with compost and manure.[28]

While the growing world population is increasing the pressure on land and water resources, economic growth and individual wealth are shifting people from predominantly starch-based diets to meat and dairy, which require more water. Producing 1 kilogram of rice, for example, requires about 3,500 liters of water, while 1 kilogram of beef needs some 15,000 liters.[29] (See Table 2.) This dietary shift has had the greatest impact on water consumption over the past 30 years and is likely to continue well into the middle of this century, according to FAO.[30]

Water challenges are compounded by the fact that agriculture competes with other uses, including hydropower. All forms of energy require water at some stage of their life cycle, which includes production, conversion, distribution, and use. Energy and electricity consumption are likely to increase over the next 25 years in all regions, with the majority of this increase occurring in non-OECD countries. This will have direct implications for the water resources needed to supply this energy. It is anticipated that water requirements for energy production will increase by 11.2 percent by 2050 if the current mix of energy sources is maintained.[31] Under a scenario that assumes increasing energy efficiency of consumption modes, the World Energy Council

Table 2. Water Required to Produce Selected Foods	
	Water Needed
	(liters per kilogram)
Crop	
Potato	500–1,500
Wheat	900–2,000
Alfafa	900–2,000
Corn/Maize	1,000–1,800
Sorghum	1,100–1,800
Soybeans	1,100–2,000
Rice	1,900–5,000
Animal Product	
Eggs	3,300
Chicken	3,500–5,700
Goat	4,000
Sheep	6,100
Beef	15,000–70,000

Source: Pacific Institute, "Water Content of Things," at www.worldwater.org/data20082009/Table19.pdf.

estimates that water requirements for energy production could decrease by 2.9 percent by 2050.[32]

Luckily, there are also technical solutions to more-efficient water use in the energy sector. For example, brackish water, mine pool water, or domestic wastewater and dry cooling techniques have been used for cooling power plants. Research is also ongoing into the water efficiency of biofuels, the energy efficiency of desalination, and the reduction of evaporation from reservoirs.[33]

Climate change will also affect global water resources at varying levels. Reductions in river runoff and aquifer recharge are expected in the Mediterranean basin and in the semiarid areas of the Americas, Australia, and southern Africa, affecting water availability in regions that are already water-stressed. In Asia—in particular, in countries such as Pakistan—the large areas of irrigated land that rely on snowmelt and high mountain glaciers for water will be affected by changes in runoff patterns, while highly populated deltas are at risk from a combination of reduced inflows, increased salinity, and rising sea levels. And rising temperatures will translate into increased crop water demand everywhere.[34]

To combat the effects of climate change, efforts must be made to follow an integrated water resource management approach on a global scale. This involves water management that recognizes the holistic nature of the water cycle and the importance of managing trade-offs within it, that emphasizes the importance of effective institutions, and that is inherently adaptive.[35]

Advertising Spending Continues Gradual Rebound, Driven by Growth in Internet Media

Shakuntala Makhijani

Global expenditures on advertising grew 3.3 percent in 2012 to $497.3 billion.[1] (See Figure 1.) The United States continues to account for the largest share of total spending, although its share is shrinking. U.S. advertising expenditures grew by 4.3 percent in 2012 and are still nearly a third of the global total.[2] (See Figure 2.) The Asia Pacific region accounted for the fastest growth, however, with ad spending there increasing by 7.9 percent in 2012 (excluding Japan, which grew by 3.1 percent and is measured separately as a fully industrialized economy).[3] Expenditures fell by 2.2 percent in Western Europe, the only region to see a decline, largely due to the ongoing Eurozone crisis.[4] The 2012 growth

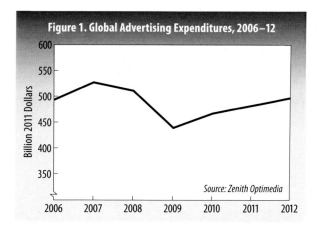

Figure 1. Global Advertising Expenditures, 2006–12

Source: Zenith Optimedia

continues the gradual rebound since advertising spending worldwide dropped by a sudden 9.6 percent in 2009 as a result of the global economic downturn.[5]

Advertising spending has responded to shifts in popular media. Internet advertising was the fastest-growing sector in 2012 and now accounts for 18 percent of the total.[6] The growth in spending on Internet ads has been driven by the expansion of social media and online video advertising.[7] Mobile and social media now account for more than half of all advertising revenue in the United States, for example, having increased by more than 30 percent in both 2011 and 2012.[8]

The growing share of Internet advertising has been accompanied by significant declines in print advertising. Over the past decade, the importance of newspaper advertising in particular has declined significantly, dropping from nearly a third of all expenditures in 2002 to less than a fifth in 2012.[9] (See Figure 3.) Meanwhile, the expansion of television's share of global advertising has leveled off after decades of growth: it rose from 36 percent to 40 percent of advertising expenditures between 2000 and 2012.[10]

As consumers grow overexposed to advertising, traditional forms such as television commercials, print advertising, and billboards are becoming less effective. As a result, advertisers are turning to subtler techniques, such as promotional material on blogs, product placement, and interactive advertising on social media such as Facebook and Twitter.[11] The distinction between advertising and media content is therefore increasingly blurred.[12] For example, global product placement expenditures are increasing rapidly, reaching $8.2 billion in 2012.[13] The United

Shakuntala Makhijani is a research associate in the Climate and Energy Program at Worldwatch Institute.

Figure 2. Global Advertising Expenditures by Region, 2012

Western Europe 21%
Latin America, 8%
Asia Pacific 28%
Central & Eastern Europe, 5%
Middle East & North Africa, 1%
Rest of World, 2%
North America 35%

Source: Zenith Optimedia

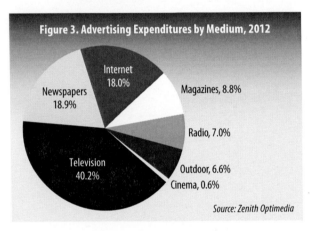

Figure 3. Advertising Expenditures by Medium, 2012

Internet 18.0%
Newspapers 18.9%
Magazines, 8.8%
Radio, 7.0%
Television 40.2%
Outdoor, 6.6%
Cinema, 0.6%

Source: Zenith Optimedia

States accounts for more than half of the product placement market worldwide, but China's market is the fastest-growing one.[14]

Advertising expenditures in the United States, the world's largest ad market, were up in 2012 across all sectors except financial services and insurance companies.[15] Retail companies account for nearly one fifth of total advertising spending, followed closely by the automobile industry.[16] (See Figure 4.)

The impacts of advertising and consumerism on all aspects of society and culture—from food choices to young girls' self-image—are well documented. Advertising targeted at children is particularly penetrating and influential, defining their identity as consumers from an early age and interfering with normal childhood development.[17] Evidence has shown that children are experiencing increased physical, emotional, and social harm as a result of consumerism through advertising.[18]

The United States is the world's largest ad market, and as a result the social impacts of advertising are studied more thoroughly there than in other countries. U.S. consumer advocates continue to call for limits on the extent and influence of advertising, especially in environments such as health clinics and public spaces as well as advertising specifically targeted at children. In particular, advertising in public schools has gained force in recent years and has infiltrated nearly all aspects of student life. Examples of this include free book covers featuring Kellogg's Pop Tarts and Fox TV characters, a nutrition curriculum provided by the Hershey Corporation, a business course curriculum from McDonald's that gives students instructions for applying for a job at the fast-food chain, and a video on environmental issues produced by Shell Oil.[19]

Due to public education spending cuts across the United States, many states are considering removing bans on school bus advertising in order to fill the budget shortfall despite the relatively low revenue generated in school districts already using bus ads.[20] Organizations including the Campaign for a Commercial-Free Childhood and Public Citizen's Commercial Alert program have recently launched efforts to prevent advertising on public school buses, including from fast-food corporations such as Burger King and Wendy's.[21]

There is extensive evidence of the negative health impacts of food and drug advertising on consumers. Fast-food and soda advertising have been found to contribute to poor diets and associated health risks, especially among children.[22] With regard to medical advertising, the rapid expansion of direct-to-consumer advertising of prescription medicines over the past decade in the United States has put pressure on the traditional physician-patient relationship. Prescriptions for heavily advertised drugs have increased much more rapidly than those for other medicines, resulting in increased overall prescribing, prescribing shifts toward less effective drugs, and increased prescribing of costlier drugs.[23] The United States and New Zealand are the only countries that allow direct-to-consumer pharmaceutical advertising.[24]

In addition to general societal advertising impacts, recent studies have shown that advertising for alcohol and tobacco are especially targeted at low-income and minority communities. Researchers at Harvard found that a low-income, minority community they studied had more tobacco retailers than a high-income, non-minority community they assessed.[25] Furthermore, these cigarette retailers were more likely to have large advertisements, promote menthol cigarettes, and be located near schools, thus promoting cigarette consumption by minors.[26] Similarly, alcohol advertising is more prominent in black and Hispanic neighborhoods.[27]

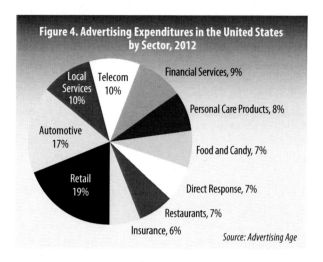

Figure 4. Advertising Expenditures in the United States by Sector, 2012

Local Services 10%
Telecom 10%
Financial Services, 9%
Personal Care Products, 8%
Automotive 17%
Food and Candy, 7%
Retail 19%
Direct Response, 7%
Restaurants, 7%
Insurance, 6%

Source: Advertising Age

Advertising continues to reinforce gender stereotypes. Women are far more likely than men to be portrayed engaging in domestic activities such as household chores and child care.[28] A comparative study across seven countries found that men were more commonly featured in prominent roles across all product advertising, while women were more often featured in ads for products typically associated with women.[29] Notably, gender differentiation in advertising was higher in the western industrial countries studied (Canada, Germany, and the United States) than in Brazil, China, South Korea, and Thailand.[30]

Advertisers have also focused more resources recently on "green" advertising aimed at attracting consumers with claims of improved environmental impact by tapping into growing public interest in sustainability and the environment. The number of new products marketed with environmental claims each year in the United States grew from around 100 in 2004 to over 1,500 in 2009.[31] The economic slump has affected the focus on green marketing, however—for example, Clorox's ad spending for its Green Works product line plummeted from over $25 million in both 2008 and 2009 to just $1.4 million in 2010.[32]

Due to increasing false claims by advertisers about product sustainability, the U.S. Federal Trade Commission established updated Green Guides in 2012 that will allow it to take enforcement action against deceptive marketing.[33] The guidelines discourage the use of general and unsubstantiated terms such as "eco-friendly" and include strong guidelines for use of terms such as "recyclable."[34]

While regulatory controls on false advertising such as the Green Guides are a positive development, true sustainability will ultimately require less material consumption and therefore stronger overall limits on advertising to stem its global growth and increasing presence in everyday life.

Population and Society Trends

Their livestock lost to drought, this family has moved to a refugee camp on the edge of Burao, Somalia

For additional population and society trends, go to vitalsigns.worldwatch.org.

Emerging Co-operatives

Gary Gardner

Approximately 1 billion people in 96 countries now belong to a co-operative—a form of business characterized by democratic ownership and governance—according to the International Co-operative Alliance.[1] Co-operatives are low-profile but powerful economic actors, with the world's 300 largest ones generating revenues in 2008 of more than $1.6 trillion.[2] If these businesses were a national economy, they would rank ninth in the world—ahead of the economy of Spain.[3]

Co-operatives, often called co-ops, are an alternative to the shareholder model of business ownership. Co-ops are governed by their members, who typically invest in the co-operative and have an ownership stake in it, as well as a voice in how the firm is run, usually on a one-member, one-vote basis. While democratic and egalitarian in outlook, most co-operatives operate in market economies and are subject to the competitive pressures found in market systems.

Co-op members use their collective power to advance their group interests.[4] Members of a worker co-op, for example, might set work hours and wage rates as well as determine when and how the firm could expand operations. Members of a consumer co-operative use their collective purchasing power to get favorable terms for their purchases. Producer co-operatives, often in the agricultural sector, seek to secure strong prices for their goods, while purchasing co-operatives are businesses that buy supplies for use by their own member businesses. Particularly in industrial countries, consumer co-ops vastly outnumber other types, accounting for 92 percent of all co-ops in the United States, for example.[5] (See Table 1.)

Co-operatives are also categorized by economic sector. At the global level, 29 percent of the 300 largest co-operatives worldwide are agricultural co-ops, 26 percent are banks and credit unions, and 22 percent are retailers and wholesalers.[6] Other sectors represented include insurance, manufacturing, health, and utilities.[7] (See Table 2.)

Although less well known than mainstream corporations, the co-operative model is widespread and by at least one measure surpasses shareholder corporations: the 1 billion member-owners of co-operatives worldwide exceed the 893 million shareholders of corporations.[8] The latter figure includes direct shareholders, who own stock as individuals, and indirect shareholders, who own stock through

Co-operative Type	Number of Co-ops	Share of Co-ops
		(percent)
Worker	223	1
Producer	1,494	5
Purchasing	724	2
Consumer	26,844	92
Total	29,285	100

Table 1. Types of Co-operatives in the United States

Source: University of Wisconsin Center for Co-operatives, "Co-operatives in the U.S. Economy," section of "Research on the Economic Impact of Co-operatives," at reic.uwcc .wisc.edu/issues.

Gary Gardner is director of grants administration at the Resources Legacy Fund in California.

mutual funds and other indirect vehicles.[9] If individual stockholders are considered alone, co-operative member-owners outnumber direct shareholders three to one.[10] Member-owner dominance of business ownership is particularly strong in Latin America.[11] (See Table 3.)

The Global 300 major co-operatives and mutual businesses are located in 25 countries, largely in the industrial world.[12] But an estimated 7 percent of Africans belong to a co-operative, and their numbers are growing rapidly.[13] The number of co-ops registered in Uganda, for example, grew 13-fold between 1995 and 2008—from 554 to nearly 7,500.[14] Savings and credit co-operatives in particular are thriving in Africa.[15]

In many countries, co-operatives have a high societal profile. In some advanced industrial countries, co-ops generate a meaningful share of economic output: 21 percent in Finland, 17.5 percent in New Zealand, 16.4 percent in Switzerland, and 13 percent in Sweden.[16] And in some countries, a sizable share of the population—up to 70 percent—belongs to a co-operative of one sort or another.[17] (See Figure 1.)

Co-operatives are particularly strongly represented in the financial realm. A 2010 World Bank report found that credit union branches account for 23 percent of bank branches worldwide and serve 870 million people, making them the second largest financial services network in the world.[18] And in several countries—Austria, Burundi, Germany, Hungary, South Korea, and Spain—branches of co-operative banks outnumber those of commercial banks.[19] Similarly for insurance: more than 30 percent of the market in the five largest insurance markets globally is controlled by mutual and co-op firms.[20]

Table 2. Top Seven Economic Sectors of Co-operative Businesses

Sector	Revenue	Share of All Co-operatives
	(billion dollars)	(percent)
Agriculture/Forestry	472	29
Banking/Credit Unions	430	26
Consumer/Retail	354	22
Insurance	282	17
Workers/Industrial	35	2
Health	27	2
Utilities	18	1
Other	17	1
Total	1,600	100

Source: International Co-operative Alliance, Volume of International Cooperation, Vol. 100, No. 1 (Geneva: 2007).

Table 3. Extent of Various Forms of Business Ownership, by Region

	Population Who Are		
Region	Co-operative Members	Indirect Shareholders	Direct Shareholders
	(percent)		
Africa	7.4	4.1	1.3
Americas	19.4	16.7	9.2
Asia Pacific	13.8	6.9	4.4
Europe	16.0	12.9	7.5
World	13.8	8.7	5.0

Source: Ed Mayo, Global Business Ownership 2012: Members and Shareholders Across the World (Manchester, U.K.: Co-operatives UK, 2012).

Nevertheless, the number of credit unions in most regions was flat or declining between 2006 and 2010, largely because of the global recession.[21] (See Figure 2.) The major exception is Africa, where the number of credit unions showed strong growth after 2006 before slumping in 2008 and then rebounding to nearly pre-recession levels by 2010.[22] In Africa, the ranks of credit unions more than doubled in the 2006–10 period and helped push growth at the global level into positive territory during those years.[23]

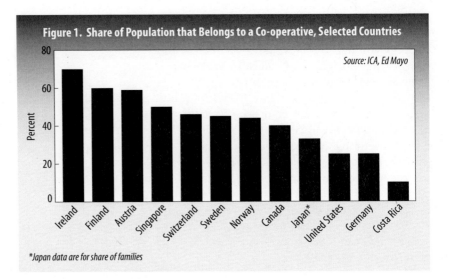

Figure 1. Share of Population that Belongs to a Co-operative, Selected Countries

Source: ICA, Ed Mayo

*Japan data are for share of families

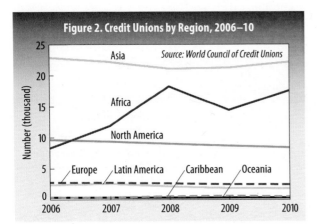

Figure 2. Credit Unions by Region, 2006–10

Source: World Council of Credit Unions

Co-operatives in the financial sector are valuable because they offer small and medium enterprises access to savings, loans, and other financial services, particularly in rural areas, which lowers a key barrier to economic development in the poorest sectors of many economies.[24] Some 45 percent of the branches of co-ops are located in rural areas, for example, compared with 26 percent of branches of commercial banks.[25]

Co-operatives may be a resilient business model that is better positioned to ride out a down economy than investor-owned businesses are. A 2009 report from the International Labour Organization notes that member-owned businesses tend to be more risk-averse than other firms.[26] Without the pressure to generate ever-higher profits for shareholders, co-operative savings and credit institutions are not compelled to devise high-risk financial products to generate greater returns for shareholders.[27] Moreover, co-operative financial institutions are funded through member deposits, limiting their need to turn to capital markets for funding.

Co-operatives were increasingly in the spotlight in 2012, which the United Nations designated the International Year of Co-operatives.[28] Governments and the civil sectors of many countries were active in promoting the sector. In the United States, for example, the National Co-operative Development Act was introduced in Congress in late 2011 to stimulate further development of the co-operative sector by providing it with seed capital, training, and resources.[29]

Climate Change Migration Often Short-Distance and Circular

Lori Hunter

Recent reports suggest that climate change, and in particular sea level rise, may be occurring faster than earlier anticipated.[1] This has increased policy and public discussions as to climate change's likely impacts on population movements, both internal and international. Traditional understandings of migration fall increasingly short of integrating the panoply of reasons why people now decide to move.

The 1951 U.N. Convention Relating to the Status of Refugees, as amended by a 1967 Protocol, defines a refugee as a person who left his or her country "owing to a well-founded fear of being persecuted for reasons of race, religion, nationality, membership of a particular social group, or political opinion."[2] Climate change is a factor more recently shaping migration streams, although its impact was not evident when the convention was drafted or amended.

The question of climate change migration needs to be seen against the backdrop of existing voluntary and involuntary population movements, which may be as high as 1 billion, according to U.N. Development Programme estimates.[3] Long-term international migrants (people who leave their home country for at least a year) are estimated at over 200 million.[4] And although numbers fluctuate with every new political crisis, refugees number nearly 10 million in the latest U.N. estimates.[5] Internally displaced people (who, unlike refugees, did not cross an international border) are estimated at 27 million.[6]

Recent research has added nuance to the scientific understanding of the potential connection between human migration and climate change. Within the past two decades, a variety of fairly alarmist estimates of future numbers of "environmental refugees" have been put forward. These range from 150 million to 1 billion and are often based on descriptive data and simplistic assumptions.[7] Several central issues that are important in the development of appropriate policy are masked in such broad-sweeping generalizations.

First, environmental drivers are rarely the only factor leading to migration.[8] Rainfall shortages and heat waves interact, for example, with persistent impoverishment and land degradation, as well as political and economic pressures.[9] In addition, in many regions women's inability to limit their family size, combined with the unmet demand for family planning, results in unsustainable population pressures on local natural resources.[10] Across the world, millions of individuals struggle daily with these challenges—and climate change is exacerbating their problems.[11]

Second, environmentally related migration is not new: migration has represented a livelihood strategy for millennia.[12] Consider low-lying Bangladesh, where migration has long served as an adaptive strategy. Over two thirds of Bangladeshis work in agriculture, forests, or fisheries—all livelihoods that depend on environ-

Lori Hunter is associate professor of sociology and environmental studies and associate director of the University of Colorado Population Center in the Institute of Behavioral Science at the University of Colorado Boulder.

mental conditions.[13] In addition, natural disasters plague rural Bangladesh with regular exposure to flooding as well as crop failure due to rainfall deficits. Food insecurity abounds.

Given these environmental challenges, and the seasonality of many occupations, circular migration from rural Bangladesh has long been a household strategy. Typically, one or two male household members move to nearby urban areas for six months of work in mills or construction or pulling rickshaws.[14]

Researchers Clark Gray and Valerie Mueller argue that the "conventional view of disaster-induced migration"—predicting millions of refugees fleeing rural Bangladesh—"is in need of considerable revision."[15] When flooding reaches higher-than-typical levels, migration may actually decline since people lack the resources to move. In addition, long-distance relocation may not be in their best interest, since movement can actually remove people from regions receiving relief assistance.[16] Instead, most environmentally related migration appears to be short-term and cyclical and represents adaptation to ongoing environmental challenges.[17]

Also intersecting with environmental factors to "push" migrants in Bangladesh are population pressures. Development scholar Katha Kartiki notes that landholdings get subdivided as households expand.[18] New research from a long-term study of Matlab, Bangladesh, finds that access to family planning programs yields social, economic, and health benefits, including smaller family size, more farmland, greater investments in perennial crops, more valuable homes, and, in general, greater assets.[19] In this way, reducing population pressures enhances resilience to climate-related stresses.

As a third key point regarding climate-related migration, it is important to remember that migration comes with costs—social, financial, and otherwise. People tend to be attached to their homelands, their cultures, their way of life. Consider the residents of Tuvalu, a low-lying Pacific island highly vulnerable to sea level rise. Based on interviews with island residents, researchers Colette Mortreux and Jon Barnett conclude that people want to stay on Tuvalu for reasons of lifestyle, culture, and identity.[20]

Even in disaster-prone rural Bangladesh, residents of *chars*—new land formed through accretion in the middle of rivers—are accustomed to recurring floods as part of their daily life. Researcher Haakon Lein concludes that "it is misleading to perceive the *chars* as high-risk areas filled with marginalized, poor people living on the brink of disaster."[21] Although life there is difficult, residents contend that it is also rewarding.

What is distinct in contemporary times, however, is the sheer number of households potentially involved in environmentally related migration, in combination with a dramatic lessening of viable livelihood options in many regions across the globe. A recent report by the U.N. High Commissioner for Refugees explores the migration-environment connection in "hotspots" of environmental change and demographic pressures.[22] Such "hotspots" include Mexico and Central America, where deforestation, land degradation, soil erosion, and droughts already render agriculture-based livelihoods highly vulnerable.[23] In Egypt, the most productive zones are the Nile Delta and Nile Valley—a region highly vulnerable to sea level rise and desertification.[24]

To quantify populations at risk, researchers are merging datasets to reflect population and environmental conditions. One example is the work of researcher Deborah Balk and her colleagues. Focused on urban vulnerability in Africa, Asia, and South America, they estimate the population at risk from climate change in two zones expected to experience serious impacts: low-level coastal zones, which are vulnerable to flooding and related health risks (such as cholera and diarrheal diseases), and arid drylands, where urban residents are often not adequately served by distribution systems even if water is plentiful.[25]

Using demographic data from the U.N. Population Division, the researchers note that, in Asia, over 300 million individuals live in dryland cities and over 200 million live in urban low-elevation coastal zones.[26] Figure 1 illustrates the intersection of population density and low-lying coastal zones in Bangladesh.[27] Dhaka is particularly vulnerable, with more than 10 million residents at risk of both coastal and inland flooding.[28] In all, low-elevation coastal zones cover 2 per cent of the world's land area but contain 10 percent of global population and 13 percent of the city dwellers.[29]

In addition, over half of Africa's urban residents are in cities located in arid zones vulnerable to water scarcity and lacking infrastructure and resources to improve resilience and lessen potential outmigration.[30]

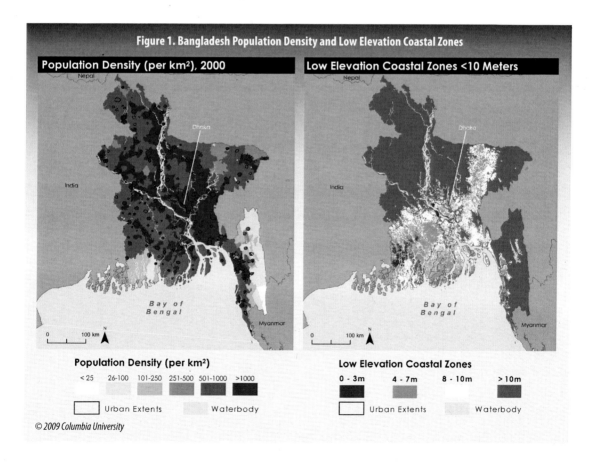

Figure 1. Bangladesh Population Density and Low Elevation Coastal Zones

© 2009 Columbia University

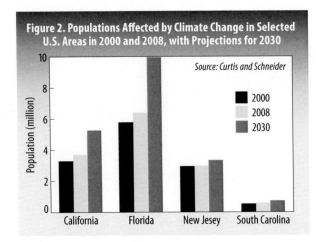

Figure 2. Populations Affected by Climate Change in Selected U.S. Areas in 2000 and 2008, with Projections for 2030

Source: Curtis and Schneider

For the United States, researchers Katherine Curtis and Anna Marie Schneider merged climate scenarios with county-level population data. They estimate that 20 million U.S. residents will be affected by sea level rise by 2030.[31] (See Figure 2.) They argue that inland regions also stand to be affected by climate change as coastal migrants relocate.[32]

In all, climate pressures on human migration cannot be denied. Yet new data and research are shedding light on the complexities underlying migration decisionmaking, as well as providing more precise estimates of vulnerable populations. Future estimates of potential climate-related migration must take these insights into account.

The key messages are that environmental change interacts with existing challenges, including persistent impoverishment, unsustainable livelihoods, and population pressures. Such challenges can impel relocation in some cases but constrain it in others. When climate-related movement does occur, research suggests much of it will be short-distance and within national borders, as opposed to international. And like much environmentally influenced migration, people's movements may be cyclical as opposed to permanent. Distinct, however, from the past is the number of individuals who could be affected and the desperation brought on by a lack of viable livelihood alternatives.

Urbanizing the Developing World

Grant Potter

Census data in 2010 indicate that cities are home to 3.5 billion people, which is 50.5 percent of the world's population.[1] Only two centuries ago, humans were predominately rural dwellers, with just 3 percent of us living in cities.[2] According to U.N. estimates, the balance tipped sometime in 2008, when more people lived in urban areas than in rural communities—a first in the history of humanity.[3]

This trend of urban population growth outpacing rural growth is expected to intensify in the future. The U.N. Population Division projects that between 2011 and 2050, the world's population will increase by 2.3 billion people, bringing the total population to 9.3 billion (the mid-level estimate).[4] During those years, ever-increasing urban populations are projected to grow by 2.6 billion people, bringing the number of urbanites to 6.3 billion.[5] Thus in the next 40 years, new and existing cities will have to cope with all the additional 2.3 billion people on Earth as a result of natural increase plus an extra 300 million people who move there from rural communities.[6]

This expected staggering growth in urban populations is likely primarily to affect developing countries. The industrial world has little room to urbanize further: it was 78 percent urban in 2011, and by 2050 it is expected to be approximately 86 percent urban.[7] (Many cities in industrial countries could continue to grow in overall population as national populations continue to rise, however, even if their proportion of the national population stays stable.) In comparison, the developing world was only 47 percent urban in 2011, and by 2050 the figure could reach 64 percent.[8] (See Figure 1.)

Given that 82 percent of the world lives in developing countries, every percentage point of urban growth there corresponds to a much higher number of people in absolute terms.[9] For example, while the developing world is less urbanized than the industrial world in relative terms, there are nonetheless 1.54 billion more people living in developing-world cities than in industrial-world cities.[10] In absolute terms, the developing world is projected to add approximately 2.45 billion people to its cities by 2050 while the industrial world is due to add just 170 million.[11] (See Figure 2.)

Within the developing world, the two regions that will almost certainly see the vast majority of this urban growth are Asia and Africa. Asia far outstrips Africa in total population, with 4.2 billion people in 2011 compared with Africa's 1 billion.[12] But these regions are also the least urbanized areas on Earth: Asia's population was

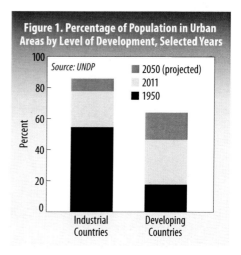

Figure 1. Percentage of Population in Urban Areas by Level of Development, Selected Years

Source: UNDP

- 2050 (projected)
- 2011
- 1950

Grant Potter is a development associate at Worldwatch Institute.

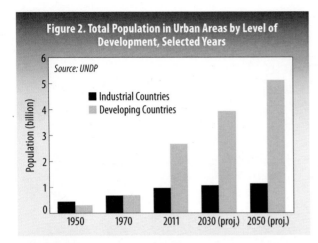

Figure 2. Total Population in Urban Areas by Level of Development, Selected Years

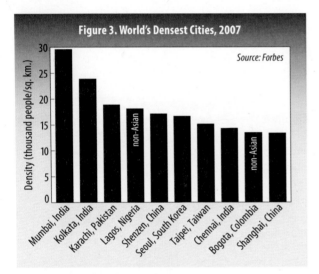

Figure 3. World's Densest Cities, 2007

45 percent urban in 2011 and Africa's only 40 percent.[13] In Latin America and the Caribbean, by contrast, 78 percent of the region's 599 million people live in cities.[14]

Africa is projected to become a majority urban region by 2035.[15] According to the U.N. Department of Economic and Social Affairs, the continent's urban population will grow from approximately 414 million to 1.26 billion between 2010 and 2050.[16]

Asia has undergone a huge urban transformation in recent history. According to the Asian Development Bank, the region is expected to have a majority urban population by 2025, a full 10 years before Africa does, and it will add another billion people by 2040.[17] What is most remarkable about this is the speed at which Asia underwent urbanization. It took Europe 150 years to go from 10 percent to over 50 percent urban, something that Asia is expected to undergo in only 95 years despite its massive size.[18]

A characteristic feature of Asian urbanization is the prevalence of megacities, areas that are home to more than 10 million people. In 2011, there were 23 such cities worldwide, 13 of which were Asian.[19] (See Table 1.) By 2025, the total number of megacities is expected to reach 37—with 21 megacities in Asia alone.[20] In 2007, Asia was home to 8 of the 10 most densely populated cities in the world.[21] (See Figure 3.) Southeast Asia is the most densely settled subregion in Asia, with approximately 16,500 people per square kilometer (compared with only 4,345 per square kilometer in Europe in 2000).[22]

Cities, especially those in the developing world, must find a way to provide essential services to their ever-increasing populations. When cities fail to meet these essential needs on a large scale, they create areas recognized world over as slums. Slum households are those that lack safe drinking water, safe sanitation, a durable living space, or security of a lease.[23] According to UN-HABITAT, 828 million people in developing-world cities are considered slum dwellers—one in every three residents.[24] (See Figure 4.) Slum populations are expected to grow significantly as world and urban populations expand in the future. In spite of some successful slum reduction programs, UN-HABITAT projects that 6 million more people become slum dwellers every year.[25]

Life in the slums is harsh. The World Health Organization (WHO) identifies

Rank	City	Population	Region
Table 1. World's 23 Megacities, 2011			
		(million)	
1	Tokyo, Japan	37.2	Asia
2	Delhi, India	22.7	Asia
3	Mexico City, Mexico	20.4	Latin America
4	New York-Newark, United States	20.4	North America
5	Shanghai, China	20.2	Asia
6	São Paulo, Brazil	19.9	Latin America
7	Mumbai, India	19.7	Asia
8	Beijing, China	15.6	Asia
9	Dhaka, Bangladesh	15.4	Asia
10	Kolkata, India	14.4	Asia
11	Karachi, Pakistan	13.9	Asia
12	Buenos Aires, Argentina	13.5	Latin America
13	Los Angeles, United States	13.4	North America
14	Rio de Janeiro, Brazil	12.0	Latin America
15	Manila, Philippines	11.9	Asia
16	Moscow, Russia	11.6	Asia
17	Osaka-Kobe, Japan	11.5	Asia
18	Istanbul, Turkey	11.3	Europe
19	Lagos, Nigeria	11.2	Africa
20	Cairo, Egypt	11.2	Africa
21	Guangzhou, China	10.8	Asia
22	Shenzhen, China	10.6	Asia
23	Paris, France	10.6	Europe

Source: U.N. Population Division, World Urbanization Prospects 2011 *(New York: 2012).*

the rapid increase of urban populations, especially slum populations, as the most important issue affecting health in the twenty-first century.[26] Overcrowding, lack of safe water, and improper sanitation systems are cited by WHO as the primary factors that contribute to poor health among the urban poor.[27] For example, over 40 percent of urban residents in sub-Saharan Africa and more than half of those in South Asia lack access to sanitation services.[28]

Combine poor sanitation with a highly concentrated population, and slums become breeding grounds for diseases like tuberculosis, dengue, pneumonia, and cholera.[29] Slum dwellers contract water-borne or respiratory illnesses at much higher rates than people in rural areas do.[30] Close to half of the city dwellers in Asia, Africa, and Latin America suffer from at least one disease that is attributed to improper sanitation or impure drinking water.[31]

Yet despite the hardships of the slums, people appear to prefer to live in them in-

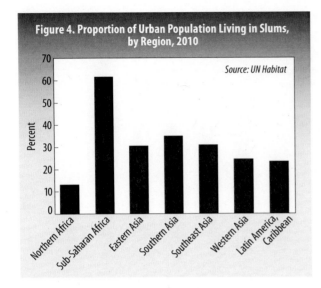

Figure 4. Proportion of Urban Population Living in Slums, by Region, 2010

Source: UN Habitat

stead of moving back to the countryside.[32] Just 25 percent of the people who live on a dollar a day live in cities; the rest are rural inhabitants. According to UN-HABITAT, poverty is 60 percent more likely in rural than in urban areas.[33] UN-HABITAT acknowledges that moving to a slum represents "the first step out of rural poverty" because of the possibility for better employment and access to health care and education services not found in rural environments.[34]

And slums do not always remain slums. Between 2000 and 2010, the sanitation and water access for 227 million people was improved to the point where they are no longer considered slum dwellers.[35] These improvements made to former slums exceeded the 2010 Millennium Development Goal targets for slums by 2.2 times.[36] Cities and their slums will continue to grow as long as rural migrants keep finding more economic and other opportunities—such as access to cultural amenities, education, and health care—in cities than in rural areas.

U.N. Funding Increases,
But Falls Short of Global Tasks

Michael Renner and James Paul

Governments have tasked the United Nations with a growing number of global mandates, but they have provided it with very few resources to carry out the work. U.N. funding is minuscule in contrast with that of other public bodies. The regular budget of the organization—$2.2 billion in 2011—is less than the total annual spending of the Tokyo Fire Department.[1] The United Nations' host city, New York, had a 2011 budget of $66 billion, about 30 times bigger than U.N. core outlays.[2] The small U.N. budget is striking in view of the multiplying global crises that need commonly decided international solutions—including climate change, financial instability, resource limits, transborder disease, and poverty.

The U.N. core "regular" budget, funded by mandatory national assessments, covers many different costs—meeting expenses, staff salaries, building maintenance, travel, security, conflict mediation, development initiatives, human rights activities, and much more. That budget is down from a peak of $2.5 billion in 2009.[3] In nominal terms, it has grown almost 14-fold over the past four decades, from $157 million in 1971.[4] (See Figure 1.) When adjusted for inflation, however, the increase is just threefold.[5] This is not nearly enough to keep up with the organization's expanded membership—which stood at 132 in 1971 and is now at 193—or with multiplying program mandates.[6] The upsurge in the past decade was preceded by flat or negative trends in constant-dollar terms during much of the 1980s and 1990s, when hostility or indifference toward the United Nations in Washington provided little opportunity for budget growth.[7]

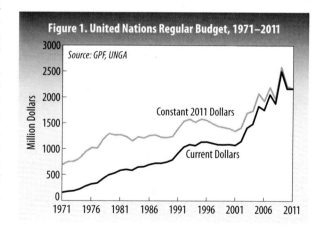

Figure 1. United Nations Regular Budget, 1971–2011

Source: GPF, UNGA

Beyond the core U.N. budget is the much larger peacekeeping budget, also financed through mandatory national assessments. These assessments include a premium paid by the five permanent Security Council members: China, France, Russia, the United Kingdom, and the United States.[8] The peacekeeping budget rises and falls according to the number and size of missions mandated by the Security Council. These missions rose steadily from 2001 to 2010, leading to considerable cost increases, as the United Nations deployed troops, police, and other peacekeeping personnel in more than a dozen crisis zones. In the budget year that spans 2011–12, the peacekeeping outlay was $7.84 billion, an expenditure that sustained more than 100,000 personnel in the field as well as logistics, equipment,

Michael Renner is a senior researcher at Worldwatch Institute. James Paul is the former executive director of the Global Policy Forum in New York City.

and headquarters staff.[9] But by comparison, individual states' military spending in 2010 (the most recent year for which estimates are available), was $1,630 billion—about 208 times larger.[10] Governments also field about 200 times as many soldiers as there are U.N. peacekeepers.[11]

Separate budgets finance the specialized U.N. agencies, programs, funds, and other entities, including the World Health Organization, UNICEF, and the U.N. Development Programme, among others. Most of these organizations function autonomously and have their own governing boards.[12] Many are located outside New York—in Geneva, Vienna, Nairobi, and other cities. Some, like UN Women and UN AIDS, are of quite recent origin; others, like the International Labour Organization, were set up several decades before the United Nations itself.[13] In 2011, the funding available to the 32 most important agencies amounted to roughly $20 billion.[14] Combined with peacekeeping spending and outlays under the core budget, the grand total for the U.N. system overall is thus about $30 billion.

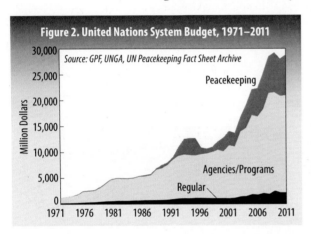

Figure 2. United Nations System Budget, 1971–2011

Source: GPF, UNGA, UN Peacekeeping Fact Sheet Archive

Peacekeeping

Agencies/Programs

Regular

U.N. funding over time shows considerable nominal growth, particularly for the various agencies and programs and for peacekeeping in more recent years.[15] (See Figure 2.) Because of gaps in available data, the figure includes the budgets of only about two dozen agencies and thus is somewhat lower overall than $30 billion in 2011. This expansion (shown in current dollars) needs to be seen against the multiplying tasks and expectations that are being placed on the U.N. system. Taking inflation into account, today's budgets are about 5.5 times larger than the funds available in the early 1970s.[16]

Funding flows for the U.N. system are complex to map. Some entities are supported only by governments' voluntary contributions; others, by a mix of voluntary and assessed contributions. By their nature, voluntary funds do not provide steady support, and they bring inevitable policy pressures from donor countries.[17] Assessed contributions are calculated on the basis of member countries' national incomes and are in principle more reliable. In reality, however, some members pay late or take deductions from what they owe. The United States pays the largest assessed contributions, because it has the largest economy, but it also has been a consistently late payer, running up very large "arrears." At the end of May 2011, Washington owed the United Nations $1.3 billion for regular and peacekeeping budgets, which was 42 percent of the total arrears for all member states.[18]

Since 2003, assessed contributions have accounted for 14–18 percent of total funds for U.N. agencies and programs.[19] This represents a decline from the 20–25 percent range that held in the 1970s, 1980s, and 1990s.[20] There is also a trend toward U.N. system funding from private sources, such as foundations and businesses. Many governments and experts are critical of this trend toward "extra-budgetary resources," which introduces influences that do not reflect the regular governance process and the political decisions taken by the majority of nations.[21]

Trust funds, pooled resources, and in-kind contributions complicate the financial picture still further. In addition to regular monies, the United Nations appeals for supplementary funds for special purposes. The Capital Master Plan for the total refurbishment of U.N. headquarters in New York has an assessment-based budget of $1.88 billion for work that will be completed in 2015.[22] Every year, the United Nations asks for additional funding to provide emergency assistance in disasters, conflicts, or other unanticipated situations. A growing number of "public-private partnerships" involve joint projects with businesses and other private actors.[23]

Table 1 shows U.N.-related organizations with the largest individual budgets. The biggest functional areas are food and agriculture, children, and health. Environmental concerns attract relatively small amounts—some $340 million in 2011 in the budgets of the U.N. Environment Programme (UNEP), the U.N. Framework Convention on Climate Change, and the World Meteorological Organization.[24] When UNEP was founded in 1972, it was expected to play a lead role in coordinating global environmental activities. But efforts to establish a strong U.N. presence in the environmental area over the years have not met with success.[25]

The budgets of the United Nations system, though modest by national standards, are subject to heated negotiations that reflect widely different member-state priorities. The poor countries, voting in the G-77 bloc, favor more U.N. activity in social and economic fields, while the rich countries prefer an emphasis on peacekeeping. The drift away from assessed payments and toward voluntary contributions and extra-budgetary resources reflects the rich-country preference for agenda-setting through bilateral pressure rather than subjecting budgets to majority votes.[26] In this way, U.N. finance is a reflection of a world divided between countries of vastly different resources, priorities, and global aspirations. At the same time, the underfunding of the overall system reflects preferences by powerful states to go their own way, outside the multilateral system and all its inevitable constraints.

Table 1. U.N. Organizations with Largest Funding, 2011

Organization	Contributions to the Budget		
	Assessed	Voluntary	Total
		(million dollars)	
U.N. Development Programme	—	4,077	4,077
World Food Programme[1]	—	3,750	3,750
U.N. High Commissioner for Refugees	—	2,133	2,133
UNICEF	—	1,970	1,970
World Health Organization	473	1,235	1,708
Food and Agriculture Organization	504	623	1,127
International Labour Organization	363	265	628
UNESCO	327	231	558

[1] Projected. All figures rounded to nearest million.
Source: U.N. General Assembly, "Budgetary and Financial Situation of the Organizations of the United Nations System," 3 August 2010; individual agency websites.

Notes

Growth in Global Oil Market Slows (pages 2–5)

1. BP, *Statistical Review of World Energy 2012: Historical Data*, available at www.bp.com, viewed 18 June 2012.

2. Ibid.

3. Ibid.

4. Ibid.

5. Ibid.

6. Ibid.

7. Ibid.

8. Ibid.

9. Ibid.

10. Ibid.

11. Ibid.

12. Ibid.

13. Ibid.

14. Ibid.

15. Ibid.

16. Wael Mahdi, "Saudis May Not Pump Intended 10 Million Barrels after IEA Move," *Bloomberg*, 28 June 2011; Rowena Mason, "OPEC Split Threatens Increase in Saudi Oil Production," (London) *Daily Telegraph*, 29 June 2011; Benoit Faucon, "Iran Says Oil Output to Spark Price War If Not Matched by Demand," *Wall Street Journal*, 7 July 2011.

17. Campbell Robertson, "Search Continues After Oil Rig Blast," *New York Times*, 21 April 2010.

18. International Energy Agency (IEA), *Medium-Term Oil and Gas Markets 2011* (Paris: 2011), p. 67.

19. Ibid.; U.S. Department of the Interior, Bureau of Ocean Energy Management, Regulation and Enforcement, "Regulatory Reform," at www.boemre.gov/reforms.htm.

20. IEA, op. cit. note 18.

21. Ibid.

22. Ibid.

23. Ibid.

24. U.S. Department of Energy (DOE), Energy Information Administration (EIA), "Cushing, OK WTI Spot Price FOB (Dollars per Barrel)," at www.eia.gov/dnav/pet/hist/LeafHandler.ashx?n=PET&s=RWTC&f=D, and "Annual Cushing, OK WTI Spot Price FOB (Dollars per Barrel)," at www.eia.gov/dnav/pet/hist/LeafHandler.ashx?n=pet&s=rwtc&f=a, both viewed 18 June 2012.

25. Ibid.

26. Worldwatch calculations based on BP, op. cit. note 1.

27. Ibid.

28. BP, op. cit. note 1.

29. Ibid.

30. Ibid.

31. Ibid.

32. Ibid.

33. Ibid.

34. Ibid.

35. Ibid.

36. Ibid.

37. Ibid.

38. Ibid.

39. "Keystone XL Pipeline Firm Offers New Plan," *San Francisco Chronicle*, 5 May 2012.

40. Ibid.; Juliet Eilperin and Steven Mufson, "Obama Administration Rejects Keystone XL Pipeline," *Washington Post*, 18 January 2012.

41. "Keystone XL Pipeline," op. cit. note 39.

42. Ibid.

43. Eilperin and Mufson, op. cit. note 40.

44. "Keystone XL Pipeline," op. cit. note 39; Eilperin and Mufson, op. cit. note 40; DOE, EIA, "Canada," at www.eia.gov/countries/cab.cfm?fips=CA, viewed 20 June 2012.

45. DOE, op. cit. note 44.

Global Coal and Natural Gas Consumption Continue to Grow (pages 6–10)

1. BP, *Statistical Review of World Energy 2012* (London: 2012).

2. Ibid.

3. International Energy Agency (IEA), *World Energy Outlook 2012* (Paris: 2012), p. 156; BP, op.cit. note 1.

4. IEA, op. cit. note 3, p. 158.

5. BP, op. cit. note 1.

6. Ibid.

7. IEA, op. cit. note 3, p. 159.

8. BP, op. cit. note 1.

9. Ibid.

10. IEA, op. cit. note 3, p. 160.

11. Ibid., p. 160.

12. BP, op. cit. note 1.

13. Ibid.

14. IEA, op. cit. note 3, p. 158.

15. Ibid.

16. Ibid., p. 162.

17. Calculated from U.S. Energy Information Administration (EIA) statistics, *Electric Power Monthly*, 31 October 2012.

18. BP, op. cit. note 1.

19. IEA, op. cit. note 3, p. 164.

20. Ibid.

21. Ibid.

22. BP, op. cit. note 1.

23. Ibid.

24. IEA, op. cit. note 3, p. 170.

25. BP, op. cit. note 1.

26. Ibid.

27. EIA, "U.S. Coal Exports on Record Pace in 2012, Fueled by Steam Coal Growth," *Today in Energy*, 23 October 2012.

28. Ibid.

29. IEA, op. cit. note 3, p. 174.

30. Ibid., p. 175.

31. Ibid., p. 176.

32. BP, op. cit. note 1.

33. BP, "World Natural Gas Consumption Grew by 2.2%," at www.bp.com.

34. BP, op. cit. note 1.

35. Ibid.

36. Ibid.

37. Ibid.

38. Ibid.

39. Ibid.

40. Ibid.

41. Ibid. Bahrain, Iran, Iraq, Kuwait, Oman, Qatar, Saudi Arabia, Syria, United Arab Emirates, and Yemen account for these reserves in the Middle East.

42. Ibid.

43. IEA, op. cit. note 3, p. 134.

44. Ibid.

45. Ibid.

46. Ibid.

47. Ibid.

48. EIA, *Monthly Energy Review*, November 2012. To make this calculation, natural gas consumption for the first eight months in 2011 was compared with natural gas consumption for the first eight months in 2012. Data were not yet available for the latter four months in 2012, so this calculation provides the best comparison between the two years.

49. EIA, op. cit. note 48.

50. Ibid. This calculation was made by comparing the emission levels of the first eight months of 2012 with the first eight months of 2007.

51. IEA, op. cit. note 3. In 2012, oil's share of primary energy consumption was 36.5 percent in the United States. Natural gas's share was 27 percent.

52. BP, op. cit. note 1.

53. Ibid.

54. IEA, op. cit. note 3.

55. Ibid.

56. Ibid.

57. BP, op. cit. note 1.

58. Ibid.

59. Ibid.

60. Ibid.

61. Ibid.

62. IEA, op. cit. note 3, p. 148.

63. Ibid., p. 150.

China Drives Global Wind Growth (pages 11–14)

1. Global Wind Energy Council (GWEC), *Global Wind Report: Annual Market Update 2011* (Brussels: 2012).

2. Calculated from ibid.

3. Ibid.

4. Ibid.

5. Ibid.

6. Sally Bakewell, "Clean Energy Investment Rises to $260 Billion, Boosted by Solar," *Bloomberg*, 12 January 2012.

7. GWEC, op. cit note 1.

8. Ibid.

9. Ibid.

10. Chinese Wind Energy Association, *A Preliminary Survey on Limited Grid Access of Wind-generated Electricity in 2011* (Beijing: April 2012), available online in Mandarin Chinese.

11. Ibid.

12. Coco Liu, "China Rebuilds Its Power Grid as Part of Its Clean Technologies Push," *New York Times*, 20 April 2011.

13. "China: State Grid Corporation of China Profile," *Zpryme Smart Grid Insights*, March 2012.

14. Ibid.

15. GWEC, op. cit. note 1.

16. American Wind Energy Association (AWEA), *U.S. Wind Energy Fourth Quarter 2011 Market Report* (Washington, DC: 2012).

17. Ibid.

18. U.S. Energy Information Administration, "U.S. Wind Generation Increased 27% in 2011," *Today in Energy*, 12 March 2012.

19. Ibid.

20. "Tax Professionals Unfamiliar with U.S. Clean Energy Incentives, Survey Finds," *Bloomberg BNA*, 13 March 2012.

21. AWEA, "New Study: Wind Energy Success Story at Risk with 54,000 American Jobs in the Balance," press release (Washington, DC: 12 December 2011); total job figure from AWEA, "Annual Report: Wind Power Bringing Innovation, Manufacturing Back to American Industry," press release (Washington, DC: 12 April 2012).

22. European Wind Energy Association (EWEA), *Wind in Power: 2011 European Statistics* (Brussels: February 2012).

23. GWEC, op. cit. note 1.

24. Ibid.

25. EWEA, op. cit. note 22, p. 4.

26. Ibid.

27. Ibid.

28. GWEC, op. cit. note 1, p. 58.

29. Alex Morales, "Vestas Remains Top Wind Turbine Maker, Goldwind Is Second," *Bloomberg*, 25 March 2012.

30. EWEA, op. cit. note 22.

31. GWEC, op. cit. note 1.

32. Anrea Roder, "Taking India by Storm," *New Energy*, February 2012.

33. Julie Chao, "Berkeley Lab Study Shows Significantly Higher Potential for Wind Energy in India Than Previously Estimated," Lawrence Berkeley National Laboratory, Berkeley, CA, 21 March 2012.

34. GWEC, op. cit. note 1.

35. Ibid.

36. EWEA, op. cit. note 22.

37. Ibid.

38. Nicole Weinhold, "Seas the Potential," *New Energy*, February 2012.

39. Ibid.

40. Ibid.

41. Ibid.

42. Ibid.

43. Wu Qi, "China's Largest Offshore Project Now Online," *WindPower Monthly*, 5 January 2012.

44. Ibid.

45. "Chu Unveils $180m Offshore Wind Innovation Fund," *BusinessGreen*, 5 March 2012.

46. U.S. Department of Energy, *A National Offshore Wind Strategy: Creating An Offshore Wind Industry in the United States* (Washington, DC: February 2011).

47. John Archer, "Wind Turbine Growth to Slow as Market Looks East: BTM," *Reuters*, 26 March 2012.

48. Alex Morales, "Vestas Chief Unveils New 7-Megawatt Offshore Wind Turbine," *Bloomberg*, 30 March 2011.

49. Archer, op. cit. note 47.

50. Ibid.

51. REN21, *Renewables 2011 Global Status Report* (Paris: 2011), p. 21.

52. "Onshore Wind Energy to Reach Parity with Fossil-fuel Electricity by 2016," *Bloomberg New Energy Finance*, 10 November 2011.

53. Sally Bakewell, "Wind Turbine Prices Fell 4% in Second Half of '11, BNEF Says," *Bloomberg*, 6 March 2012.

54. "Onshore Wind Energy," op. cit. note 52.

Hydropower and Geothermal Growth Slows (pages 15–18)

1. Geothermal Energy Association (GEA), *Geothermal Basics, Q & A* (Washington, DC: September 2012).

2. British Petroleum (BP), *Statistical Review of World Energy June 2012*, online database, at www.bp.com/sectionbody copy.do?categoryId=7500&contentId=7068481; REN21, *Renewable 2012 Global Status Report* (Paris: 2012). Previous editions of *Vital Signs* have included a higher figure for hydropower installed capacity that included some pumped storage capacity. The 2011 figure cited here represents a change in the methodology used to calculate total capacity, which no longer includes pumped storage capacity.

3. BP, op. cit. note 2.

4. Ibid.

5. Ibid.

6. Ibid.

7. Ibid.

8. Ibid.

9. Ibid.

10. Ibid.

11. Ibid.

12. REN21, op. cit. note 2.

13. Ibid.; Ministry of Energy of the Russian Federation, "The Main Types of Electricity Production in Russia," viewed via Google Translate, January 2013.

14. REN21, op. cit. note 2.

15. Ibid.

16. Ibid.

17. Ibid.

18. International Rivers, "Wrong Climate for Big Dams," Berkeley, CA, October 2011.

19. Michael Harris, "Hydropower 2012 Year in Review: Development Around the Globe," *RenewableEnergyWorld.com*, 25 December 2012.

20. "China's Three Gorges Hydropower Plant Running at Full Capacity," *HydroWorld.com* 6 July 2012.

21. REN21, *Renewables 2011 Global Status Report* (Paris: 2011).

22. Baiba Auzāne, *Small Hydropower: Market Potential in Developing Countries* (Brussels: Alliance for Rural Electrification, 24 May 2012).

23. REN21, *Renewables 2010 Global Status Report* (Paris: 2010).

24. Auzāne, op. cit. note 22.

25. U.N. Environment Programme and Bloomberg New Energy Finance, *Global Trends in Renewable Energy Investment 2012* (Frankfurt: 2012).

26. REN21, op. cit. note 2.

27. Ibid.

28. Ibid.

29. Ibid.

30. Ibid.

31. Ibid.

32. Ibid.

33. Ibid.

34. GEA, op. cit. note 1; Ruggero Bertani, "Geothermal Power Generation in the World 2005–2010 Update Report," Proceedings of the World Geothermal Congress, 25–29 April 2010. In addition to power generation, geothermal technologies are also used to provide heat energy in 78 countries, including Iceland, which gets 90 percent of its heat through geothermal resources.

35. REN21, op. cit. note 2.

36. BP, op. cit. note 2.

37. REN21, op. cit. note 2.

38. BP, "Geothermal Capacity," in *Statistical Review of World Energy 2012*, at www.bp.com.

39. REN21, op. cit. note 2; REN21, op. cit. note 21.

40. REN21, op. cit. note 2.

41. Ibid.

42. GEA, op. cit. note 1.

43. GEA, *Geothermal: International Market Overview Report* (Washington, DC: May 2012).

44. REN21, op. cit. note 2.

45. Ibid.; REN21, op. cit. note 21.

46. Calculated from BP, op. cit. note 2.

47. REN21, op. cit. note 2.

48. I. B. Fridleifsson et al., "The Possible Role and Contribution of Geothermal Energy to the Mitigation of Climate Change," in O. Hohmeyer and T. Trittin (eds.), *IPCC Scoping Meeting on Renewable Energy Sources, Proceedings*, Luebeck, Germany, 20–25 January 2008, pp. 59–80; U.S. Department of Energy, Energy Information Administration, *Electric Power Annual 2009* (Washington, DC: April 2011).

49. Fridleifsson et al., op. cit. note 48.

50. REN21, op. cit. note 2.

Smart Grid and Energy Storage Installations Rising (pages 19–21)

1. Bloomberg New Energy Finance (BNEF), *Smart Grid Infrastructure Remains Global Growth Market*, press release (Washington, DC: 24 January 2013).

2. International Energy Agency (IEA), *Technology Roadmaps: Smart Grids* (Paris: 2011), p. 6

3. BNEF, op. cit. note 1.

4. Ibid.

5. Ibid.

6. Ibid.

7. Ibid.

8. U.S. Department of Energy (DOE), *Smart Grid Investment Grant Progress Report* (Washington, DC: 2012), p. iii.

9. Ibid., p. ii.

10. Energy Information Agency (EIA), "How Many Smart Meters Are Installed in the US and Who Has Them?" *Frequently Asked Questions*, at www.eia.gov/tools/faqs/faq.cfm?id=108&t=3, viewed 10 January 2013.

11. Edison Electric Institute (EEI), Association of Edison Illuminating Companies, and Utilities Telecom Council, *Smart Meters and Smart Meter Systems: A Metering Industry Perspective* (Washington, DC: EEI, 2011), pp. 7–8.

12. Institute for Electric Efficiency, *Utility-Scale Smart Meter Deployments, Plans, and Proposals* (Washington, DC: 2012), p. 1.

13. Zpryme Research & Consulting, *China: Rise of the Smart Grid* (Austin, TX: 2011), p. 1.

14. Ibid., p. 3.

15. Ibid.

16. Ibid.

17. Ibid, p. 1.

18. Pike Research, "Installed Base of Smart Meters in China

to Reach Nearly 380 Million by 2020," press release (Washington, DC: 15 January 2013). The percentage was calculated from the Pike Research numbers and from the number of Chinese households from United Nations, *Demographic Yearbook* (New York: 2011).

19. Calculated from Sangim Han, "South Korea's Smart Meters Program Averts Nuclear Need," *Bloomberg News*, 12 March 2012.

20. Global Smart Grid Federation (GSGF), *2012 Report* (London: 2012), p. 35.

21. Zpryme Research & Consulting, *Japan: Tsunami Wakens the Smart Grid* (Austin, TX: 2012), p. 5.

22. Rudolf ten Hoedt, "Japan Catches Up to Smart Energy," in *European Energy Review Special Report #1: The Secrets of Successful Smart Energy Approaches* (Groningen, Netherlands: 2013), p. 38.

23. Ibid., p. 39.

24. Calculated from ibid.

25. Zpryme Research & Consulting, op. cit. note 21, p. 4.

26. GSGF, op. cit. 20, p. 21.

27. Ibid.

28. European Electricity Grid Initiative, *Roadmap 2010–18 and Detailed Implementation Plan 2010–12* (25 May 2010), p. 2.

29. Enrico Giglioli, *How Europe Is Approaching the Smart Grid* (McKinsey & Company, 2010), p. 13.

30. Joint Research Centre–Institute for Energy (JRC-IE), *Smart Grid Projects in Europe: Lessons Learned and Current Developments* (Petten, Netherlands: European Commission, 2011), p. 19.

31. Ibid.

32. Giglioli, op. cit. note 29, p. 14.

33. GSGF, op. cit. note 20, p. 25.

34. Ibid.

35. Calculated from Energy Saving Trust, *Smart Meters*, at www.energysavingtrust.org.uk/Electricity/Smart-meters.

36. JRC-IE, op. cit. note 30, p. 14.

37. Rafael Figueiredo, "Brazil Update: ANEEL Finalizes the Regulatory Framework for Smart Grid Deployment," *Renewable Energy World*, 11 September 2012.

38. IEA, *Electricity Storage Technology Brief* (Paris: 2012), p. 4.

39. DOE, *Energy Storage Activities in the United States Electricity Grid* (Washington, DC: 2011), p. 2.

40. Ibid.

41. Ibid.

42. Pike Research, "More than 700 Energy Storage Projects Are Announced or Operating Around the World," press release (Washington, DC: 9 December 2012).

43. Calculated from Pike Research, "Nearly 600 Energy Storage Projects Have Been Announced or Deployed Worldwide," press release (Washington, DC: 25 October 2011).

44. Pike Research, op. cit. note 42.

45. Pike Research, *Energy Storage on the Grid: Executive Summary* (Washington, DC: 2012), p. 1.

Fossil Fuel and Renewable Energy Subsidies on the Rise (pages 22–24)

1. Oil Change International, *No Time to Waste: The Urgent Need for Transparency in Fossil Fuel Subsidies,* 15 May 2012, at priceofoil.org/wp-content/uploads/2012/05/1TFSFIN .pdf, pp. 1–2.

2. International Energy Agency (IEA), *World Energy Outlook 2011* (Paris: 2011), p. 530.

3. Ibid.

4. International Institute for Sustainable Development (IISD), *Joint Submission to the UN Conference on Sustainable Development, Rio +20: A Pledge to Phase Out Fossil-fuel Subsidies* (Winnipeg, MB: 2011), p. 10.

5. IEA, op. cit. note 2, p. 507.

6. Doug Koplow, *Measuring Energy Subsidies Using the Price-Gap Approach: What Does It Leave Out?* (Winnipeg, MB: IISD, 2009), p. 1.

7. Doug Koplow, *G20 Fossil-Fuel Subsidy Phase Out: A Review of Current Gaps and Needed Changes to Achieve Success* (Cambridge, MA: Earth Track, 2010), p. 2.

8. Doug Koplow, *Phasing Out Fossil Fuel Subsidies in the G20: A Progress Update* (Cambridge, MA: Earth Track, 2012), p. 6.

9. National Academy of Sciences, *Hidden Costs of Energy: Unpriced Consequences of Energy Production and Use* (Washington, DC: 2009).

10. Chris Nelder, "Reframing the Transportation Debate" (blog), SmartPlanet.com, 19 October 2011.

11. U.N. Environment Programme, *Universal Ownership: Why Environmental Externalities Matter to Institutional Investors* (Nairobi: Finance Initiative, 2010), p. 4.

12. IEA, *Joint Report by IEA, OPEC, OECD and World Bank on Fossil-fuel and Other Energy Subsidies: An Update of the G20 Pittsburgh and Toronto Commitments* (Paris: 2011).

13. Ibid.

14. Ibid.

15. Figure 1 is for illustration purposes only. It displays two different data sets from the IEA and the Organisation for Economic Co-operation and Development (OECD). The subsidy values listed for developing countries are from the IEA data set and are derived using the price-gap method of calculating subsidies. The industrial-country values are an aggregate of the individual country values listed by the OECD. These values were calculated using a broader definition of support (including direct budgetary transfers and tax expenditures) than is included in the price-gap method. As such, the numbers are not directly comparable and do not necessarily all reflect inefficient subsidies. However, given the lack of a clear definition of "subsidy," Figure 1 seeks to illustrate the general trend of global fossil fuel and renewable consumption subsidies.

16. Ibid.

17. Ibid.

18. Ibid.

19. OECD, "OECD-IEA Fossil Fuel Subsidies and Other Support," at www.oecd.org/site/tadffss/#data, viewed 13 August 2012.

20. Ibid.

21. Ibid.

22. IEA, op. cit. note 2, p. 515.

23. Ibid., p. 518.

24. Kerryn Lang and Peter Wooders, *A How-to Guide: Measuring Subsidies to Fossil Fuel Producers* (Winnipeg, MB: Global Subsidies Initiative and IISD, 2010), p. 2.

25. IISD, op. cit. note 4, p. 1.

26. IEA, op. cit. note 2, p. 511.

27. Ibid.

28. Ibid.

29. Ibid., p. 520.

30. Ibid.

31. Ibid.

32. Ibid.

33. Duncan Clark, "Phasing Out Fossil Fuel Subsidies 'Could Provide Half of Global Carbon Target,'" (London) *Guardian*, 19 January 2012.

34. Group of Twenty, *Leader's Statement: The Pittsburgh Summit* (Pittsburgh, PA: Group of Twenty, 2009), p. 3.

35. Koplow, op. cit. note 8, p. 2.

36. Ibid.

37. Dominique Guillame et al., *Iran—The Chronicles of the Subsidy Reform* (Washington, DC: International Monetary Fund, 2011), pp. 3, 12.

38. "Nigeria Restores Fuel Subsidy to Quell Nationwide Protests," (London) *Guardian*, 16 January 2012.

39. IEA, op. cit. note 2, p. 519. The 11 countries were Pakistan, China, Vietnam, Thailand, Indonesia, Angola, Bangladesh, India, Philippines, Sri Lanka, and South Africa.

40. IEA, op. cit. note 2.

41. International Finance Corporation, *Lighting Asia: Solar Off-Grid Lighting* (Washington, DC: 2012), p. 51.

Continued Growth in Renewable Energy Investments (pages 25–28)

1. U.N. Environment Programme (UNEP) and Bloomberg New Energy Finance (BNEF), *Global Trends in Renewable Energy Investment 2012* (Frankfurt: 2012).

2. Ibid.

3. BNEF, "Q2 2012 Clean Energy Policy & Market Briefing," 19 July 2012, at www.bnef.com/WhitePapers /view/114.

4. UNEP and BNEF, op. cit. note 1.

5. Ibid.

6. Ibid.

7. Ibid.

8. REN21, *Renewables 2011 Global Status Report* (Paris: 2011); REN21, *Renewables 2012 Global Status Report* (Paris: 2012).

9. Ibid.

10. Ibid.

11. Ibid.

12. Ibid.

13. UNEP and BNEF, op. cit. note 1.

14. Ibid.

15. Ibid.

16. REN21, *2012 Status Report*, op. cit. note 8.

17. Ibid.

18. Ibid.

19. UNEP and BNEF, op. cit. note 1.

20. Ibid.

21. REN21, *2012 Status Report*, op. cit. note 8.

22. UNEP and BNEF, op. cit. note 1.

23. International Energy Agency (IEA), *World Energy Outlook 2011*, factsheet (Paris: 2011).

24. IEA, *Tracking Clean Energy Progress* (Paris: 2012).

25. Intergovernmental Panel on Climate Change (IPCC), *IPCC Special Report on Renewable Energy Sources and Climate Change Mitigation* (Cambridge, U.K.: Cambridge University Press, 2011), p. 879.

26. IEA, *Energy for All, Financing Access for the Poor* (Paris: 2011).

27. UNEP and BNEF, op. cit. note 1.

28. Ibid.

29. Ibid.

30. Ibid.

31. Ibid.

32. Ibid.

33. IPCC, op. cit. note 25.

34. UNEP and BNEF, op. cit. note 1.

35. Ibid.

36. Ibid.

37. Ibid.

38. Ibid.

39. Ibid.

40. Ibid.

41. Ibid.

42. Ibid.

43. Ibid.

Auto Production Roars to New Records (pages 29–32)

1. Colin Couchman, IHS Automotive, London, e-mail to author, 10 August 2012.

2. Ibid.

3. PricewaterhouseCoopers (PWC), "Autofacts: Quarterly Forecast Update," January 2012, p. 4.

4. Ibid.

5. Couchman, op. cit. note 1.

6. Ibid.

7. Ibid.

8. Calculated from ibid. and from Population Reference Bureau, *2012 World Population Data Sheet* (Washington, DC: 2012).

9. Deutsche Bank Climate Change Advisors, *China's Green Move – Vehicle Electrification Ahead*, New York, 8 August 2012, p. 7.

10. Couchman, op. cit. note 1.

11. Ibid.

12. Ibid.

13. PWC, op. cite. note 3, p. 3.

14. Couchman, op. cit. note 1.

15. Ibid.

16. Ibid.

17. Ibid.

18. PWC, op. cit. note 3, p. 6.

19. Ibid., p. 7.

20. Ibid.

21. Ibid.

22. Ibid.

23. International Council on Clean Transportation, "Global Passenger Vehicle Standards Update. August 2012 Datasheet," Washington, DC.

24. Ibid.

25. Ibid.

26. Ibid.; White House, "Obama Administration Finalizes Historic 54.5 MPG Fuel Efficiency Standards," press release (Washington, DC: 28 August 2012).

27. European Federation for Transport and Environment, *How Clean Are Europe's Cars? An Analysis of Carmaker Progress Towards EU CO2 Targets in 2010* (Brussels: September 2011), p. 9.

28. Ibid.

29. Ibid., p. 12.

30. Ibid., p. 18.

31. Ibid., p. 26.

32. Ibid.

33. Calculated from U.S. Environmental Protection Agency, *Light-Duty Automotive Technology, Carbon Dioxide Emissions, and Fuel Economy Trends: 1975 Through 2011* (Washington, DC: March 2012), Table 3, "Carbon Dioxide Emissions of MY 1975 to 2011 Light Duty Vehicles," p. 23. The data are for "adjusted composite fuel economy"—covering combined city/highway laboratory drive cycles for cars and light trucks.

34. Ibid., Table 5, p. 29.

35. Ibid.

36. PWC, op. cit. note 3, p. 6.

37. Ibid.

38. Deutsche Bank Climate Change Advisors, op. cit. note 9, p. 17.

39. Ibid.

40. Ibid., pp. 15–16.

41. Ibid., pp. 8, 15.

42. Ibid.

43. Organisation for Economic Co-operation and Development (OECD), "Inland Passenger Transport," OECD Statistics, at www.stats.oecd.org, viewed 10 August 2012.

44. Ibid.

45. Ibid.; 2008 is the most recent year for which the OECD offers complete statistics for passenger kilometers driven.

46. OECD, op. cit. note 43.

47. Ibid.

48. Ibid.

49. Ibid.

50. Ibid.

51. International Transport Forum, *Transport Outlook 2012* (Paris: OECD, 2012), p. 43.

52. Ibid., p. 42.

53. OECD, op. cit. note 43.

54. Ibid.

55. Ibid.

56. Ibid.

Carbon Dioxide Emissions and Concentrations on the Rise as Kyoto Era Fades (pages 34–38)

1. C. D. Keeling et al., "Atmospheric CO_2 Concentrations (ppmv) Derived from In Situ Air Samples Collected at Mauna Lao Observatory, Hawaii," Scripps Institution of Oceanography, University of California, San Diego, February 2012.

2. P. Forster et al., "Changes in Atmospheric Constituents and in Radiative Forcing," in Intergovernmental Panel on Climate Change (IPCC), *Climate Change 2007: The Physical Science Basis—Contribution of Working Group I to the Fourth Assessment Report of the Intergovernmental Panel on Climate Change* (Geneva: 2007), p. 212.

3. J. G. J. Olivier et al., *Long-term Trend in Global CO_2 Emissions* (The Hague: PBL Netherlands Environmental Agency/

Joint Research Commission,2011), p. 3, based on data from BP, *Statistical Review of World Energy* (London: June 2011).

4. Ibid., p. 12.

5. International Energy Agency (IEA), *CO$_2$ Emissions from Fuel Combustion, Highlights* (Paris: 2011), p. 18.

6. G. R. van der Werf et al., "Global Fire Emissions and the Contribution of Deforestation, Savanna, Forest, Agricultural, and Peat Fires (1997–2009)," *Atmospheric Chemistry and Physics*, vol. 10, no. 23 (2010), pp. 11,707–35.

7. Cement and Concrete Institute, *Sustainable Concrete* (Midrand, South Africa: 2011), p. 6.

8. Olivier et al., op. cit. note 3, p. 10.

9. IEA, op. cit. note 5, p. 18.

10. Ibid.

11. Olivier et al., op. cit. note 3, p. 24.

12. Ibid., p. 10.

13. IEA, op. cit. note 5, p. 18.

14. BP, op. cit. note 3.

15. Ibid.

16. Ibid.

17. Ibid. Annex I countries are Organisation for Economic Co-operation and Development countries minus Chile, Isreal, South Korea, and Mexico, plus Ukraine, Belarus, and Russia. Emissions of Bulgaria, Croatia, Lativia, Lichtenstein, Malta, Monaco, Romania, and Ukraine are counted within emissions of the European Union.

18 Olivier et al., op. cit. note 3, p. 24.

19. Ibid., p. 25.

20. Ibid., p. 24.

21. United Nations Framework Convention on Climate Change, "Outcome of the Work of the Ad Hoc Working Group on Further Commitments for Annex I Parties under the Kyoto Protocol at its Sixteenth Session," Durban, 2011.

22. D. Seligsohn, "China at Durban: First Steps Toward A New Climate Agreement," *WRI Insights*, 16 December 2011.

23. BP, op. cit. note 3.

24. Ibid.

25. World Resources Institute, *Climate Analysis Indicators Tool*, electronic database, February 2012.

26. Ibid.

27. BP, op. cit. note 3; IEA, op. cit. note 5, p. 82.

28. Alexander Ochs and Xueying Wang, "Global and Na-

tional CO$_2$ Emissions: A Reality Check in Time for the Durban Summit," (blog) *Revolt*, Worldwatch Institute, 2 December 2011.

29. IEA, op. cit. note 5, p. 11.

30. IPCC, *Synthesis Report. Contribution of Working Groups I, II and III to the Fourth Assessment Report of the Intergovernmental Panel on Climate Change* (Geneva: 2007).

31. National Aeronautics and Space Administration, "NASA Finds 2011 Ninth Warmest Year on Record," press release (Washington, DC: 19 January 2012).

32. Ibid.

33. A. Strong et al., *Climate Science 2009–2010: Major New Discoveries* (Washington, DC: World Resources Institute, 2011).

34. Ibid., p. 4.

35. IPCC, *Special Report on Managing the Risks of Extreme Events and Disasters to Advance Climate Change Adaptation* (Geneva: 2011).

36. Mark G. New et al., "Four Degrees and Beyond: The Potential for a Global Temperature Increase of Four Degrees and Its Implications," Thematic Issue, *Philosophical Transactions of the Royal Society A*, 13 January 2011.

Carbon Capture and Storage Experiences Limited Growth in 2011 (pages 39–43)

1. Global CCS Institute, *The Global Status of CCS: 2011* (Canberra, Australia: 2011).

2. Global CCS Institute, "Status of CCS Project Database," 26 March 2012, at www.globalccsinstitute.com/publications/data/dataset/status-ccs-project-database.

3. Ibid. The Global CCS Institute defines "large" projects as those that sequester at least 800,000 tons of carbon dioxide (CO$_2$) at a coal power plant, or at least 500,000 tons of CO$_2$ for other facilities, including natural gas power plants. Its database of large integrated CCS projects includes plants in five stages of development: Identify, Evaluate, Define, Execute, and Operate. The eight plants referred to here are all in the fifth stage.

4. Global CCS Institute, op. cit. note 2; vehicle equivalence is a Worldwatch calculation based on U.S. Environmental Protection Agency, "Passenger Vehicle Conversion," at www.epa.gov/cleanenergy/energy-resources/refs.html#vehicles.

5. Global CCS Institute, op. cit. note 2.

6. Global CCS Institute, op. cit. note 1.

7. Ibid.

8. Ibid.

9. Ibid.

10. Ibid.

11. Global CCS Institute, *Global Status of Large-Scale Integrated CCS Projects: December 2011 Update* (Canberra, Australia: 2011).

12. Global CCS Institute, op. cit. note 2. The Gordon Carbon Dioxide Injection Projection (Australia) will store 3.4–4 million tons per year; the assumption here is that it will store 3.7 million tons.

13. Global CCS Institute, op. cit. note 2. This total does not include the storage capacity of the Caledonia Clean Energy Project in Scotland, as its storage volume is not specified.

14. Global CCS Institute, op. cit. note 2; global emissions equivalence is a Worldwatch calculation based on data from BP, *Statistical Review of World Energy* (London: June 2011).

15. Global CCS Institute, op. cit. note 1; Global CCS Institute, op. cit. note 2.

16. Global CCS Institute, op. cit. note 1.

17. Global CCS Institute, op. cit. note 11.

18. International Energy Agency (IEA), *Prospects for CO₂ Capture and Storage*, at www.iea.org/textbase/npsum/ccs SUM.pdf.

19. "EPA to Impose First Greenhouse Gas Limits on Power Plants," *Washington Post*, 26 March 2012.

20. Ibid.

21. IEA, *Technology Roadmap: Carbon Capture and Storage* (Paris: 2009).

22. Ibid.

23. Ibid.

24. Global CCS Institute, op. cit. note 1.

25. Global CCS Institute, op. cit. note 2.

26. Ibid.

27. IEA, op. cit. note 21.

28. Ibid.

29. Global CCS Institute, op. cit. note 1.

30. Ibid.

31. Ibid.

32. Global CCS Institute, op. cit. note 2.

33. Ibid.

34. Ibid.

35. Ibid.

36. Ibid.

37. Ibid.

38. Ibid.

39. Global CCS Institute, op. cit. note 1.

40. Ibid.

41. ZEP (European Technology Platform for Zero Emission Fossil Fuel Power Plants), *The Costs of CO₂ Capture, Transport and Storage* (Brussels: 2011).

42. Global CCS Institute, op. cit. note 2.

43. Global CCS Institute, op. cit. note 1.

44. Ibid.

45. Matthias Finkenrath, *Cost and Performance of Carbon Dioxide Capture from Power Generation, 2011*, Working Paper (Paris: IEA, 2010).

46. Ibid.

47. Intergovernmental Panel on Climate Change, *Carbon Dioxide Capture and Storage: Summary for Policymakers* (Geneva: 2005), p. 12.

48. "Carbon Capture Project Leaking in to Their Land, Couple Says," (Toronto) *Globe and Mail*, 11 January 2011.

49. "Method Puts Carbon Capture and Storage 'Leaks' to Test," *The Engineer*, 13 December 2011.

50. On increased water usage and fuel consumption, see National Energy Technology Laboratory, *Carbon Capture Approaches for Natural Gas Combined Cycle Systems* (Washington, DC: U.S. Department of Energy, 2010); Mark Little and Robert Jackson, "Potential Impacts of Leakage from Deep CO₂ Geosequestration on Overlying Freshwater Aquifers," *Environmental Science and Technology*, vol. 44 (2010), pp. 9,225–32.

51. J. P. Ciferno, "Determining Carbon Capture and Sequestration's Water Demands," *Power*, 1 March 2010.

52. Finkenrath, op. cit. note 45.

53. Global CCS Institute, "International Progress," 30 April 2012, at www.globalccsinstitute.com/publications/global -status-ccs-2010/online/33536.

54. Ibid.

55. Paige Andrews et al., *COP-17 De-Briefing: Enhancements, Decisions, and the Durban Package* (Climatico, 2012).

56. Ibid.

57. Ibid.

Global Grain Production at Record High Despite Extreme Climatic Events (pages 46–48)

1. U.N. Food and Agriculture Organization (FAO), *Food Outlook*, May 2012, p. 1. FAO includes coarse grains with maize data. All tons in this article are metric tons.

2. FAO, op. cit. note 1.

3. Ibid.

4. Ibid.; FAO, *Food Outlook*, June 2008.

5. International Grains Council, *Grain Market Report*, 26 July 2012, p. 3.

6. "Staple Foods: What Do People Eat?" in Tony Loftas, ed., *Dimensions of Need, An Atlas of Food and Agriculture* (Rome: FAO, 1995).

7. U.S. Department of Agriculture (USDA), Foreign Agricultural Service, "Diets Around the World: How the Menu Varies," material based on data and analysis of Stacey Rosen, Economic Research Service, USDA, and on the FAO AGROSTAT database, last updated 14 October 2004.

8. FAO, op. cit. note 1, p. 15.

9. Ibid., p. 4.

10. Ibid., p. 2.

11. Ibid., pp. 10–11.

12. Ibid., p. 16.

13. USDA, National Agricultural Statistics Service, Agricultural Statistics Board, *Crop Production*, 10 August 2012.

14. FAO, op. cit. note 1, p. 16.

15. Ibid.

16. Ibid., p. 22.

17. Ibid., p. 23.

18. Ibid.

19. Ibid.

20. Ibid.

21. Ibid., p. 2.

22. International Grains Council, *Grain Market Report*, 2 July 2012, p. 2.

23. FAO, op. cit. note 1, p. 10; Mark Memmott, "Drought Disasters Declared in More Counties: 1,207 Affected So Far," *National Public Radio*, 18 July 2012.

24. FAO, *FAOSTAT Statistical Database*, at faostat.fao.org; FAO, op. cit. note 1; USDA, Foreign Agricultural Service, *World Agricultural Production*, Circular Series WAP 08–11, August 2011; FAO, *Rice Market Monitor*, April 2012.

25. FAO, op. cit. note 24.

26. FAO, op. cit. note 1, p. 25.

27. Ibid.

28. Ibid.

29. International Grains Council, op. cit. note 22.

30. FAO, op. cit. note 1, p. 13.

31. Ibid.

32. U.N. International Strategy for Disaster Reduction, *Climate Change and Disaster Risk Reduction* (Geneva: 2008), as cited in U.N. Environment Programme (UNEP) Management Group, "Drylands in a Global Context," *Global Drylands: A UN System-wide Response* (Nairobi: UNEP, 20 October 11); World Food Programme from FAO Committee on Food Security, *Managing Vulnerability and Risk to Promote Better Food Security and Nutrition*, Policy Roundtable Background Paper (Rome: October 2010); Oxfam International, "54 Percent Increase in Number of People Affected by Climate Disasters by 2015 Could Overwhelm Emergency Responses," press release (Oxford: 21 April 2009).

33. FAO Committee on Food Security, op. cit. note 32, p. 5.

34. Sophia Murphy, in *Grain Reserves and the Food Price Crisis: Selected Writings from 2008–2012* (Minneapolis, MN: Institute for Agriculture and Trade Policy (IATP), June 2012), p. 3; Jim Harkness, "The 2050 Challenge to Our Global Food System," remarks at National Food Policy Conference, Consumers Federation of America, Washington, DC, October 2011.

35. Sophia Murphy, *Grain Reserves: A Smart Climate Adaptation Policy* (Minneapolis, MN: IATP, November 2010).

36. Alyssa A. Botelho and Joel Achenbach, "Drought Intensifies in Most-parched Areas of U.S.," *Washington Post*, 3 August 2012.

37. Mark Memmott, "Drought in U.S. Now Worst Since 1956; Food Prices to Spike, Economy to Suffer," *National Public Radio*, 17 July 2012.

38. Eric Holt-Giménez, interview with Ian Masters, "Geopolitics of Drought," GRAIN, 19 July 2012.

39. Anne Lowrey and Ron Nixon, "Food Prices to Rise in Wake of Severe Drought," *New York Times*, 25 July 2012.

40. Rockefeller Foundation, "African Agriculture & Climate Change Resilience," at www.rockefellerfoundation.org/what -we-do/current-work/developing-climate-change-resilience /african-agriculture-climate-change.

Disease and Drought Curb Meat Production and Consumption (pages 49–52)

1. U.N. Food and Agriculture Organization (FAO), "Meat and Meat Products," *Food Outlook*, May 2012.

2. Ibid.

3. FAO, *FAOSTAT Statistical Database*, at faostat.fao.org, updated 23 February 2012, and earlier editions.

4. FAO, op. cit. note 1.

5. Ibid.

6. Ibid.

7. FAO, op. cit. note 3.

8. Ibid.; FAO, op. cit. note 1.

9. FAO, op. cit. note 1.

10. Ibid.

11. Ibid.

12. Ibid.

13. Ibid.

14. Ibid.

15. FAO, op. cit. note 3.

16. FAO, op. cit. note 1.

17. Ibid.

18. FAO, op. cit. note 3.

19. Andrew E. Kramer, "Russia, Crippled by Drought, Bans Grain Exports," *New York Times*, 5 August 2010.

20. UN News Centre, "Winter Drought Threatens Wheat Harvest in North China Plain as Prices Soar—UN," press release (New York: 8 February 2011).

21. Ibid.

22. FAO, op. cit. note 1.

23. National Oceanic and Atmospheric Association (NOAA), *State of the Climate: Drought, Annual 2011* (Washington, DC: January 2012).

24. Texas A&M Agricultural Extension Service, "Texas Agricultural Drought Losses Reach Record $5.2 Billion," press release (College Station, TX: 17 August 2011).

25. NOAA, op. cit. note 23.

26. FAO, "Meat and Meat Products," *Food Outlook*, November 2011.

27. U.S. Department of Agriculture, "Changes in Food Price Indexes, 2010 through 2013," press release (Washington, DC: 24 August 2012).

28. FAO, *FAO Food Price Index* (Rome: 9 June 2012).

29. FAO, op. cit. note 1.

30. D. Grace et al., *Mapping of Poverty and Likely Zoonoses Hotspots: Report to the Department for International Develop-*

ment (Nairobi: International Livestock Research Institute (ILRI), 2012).

31. FAO, *Livestock's Long Shadow, Environmental Issues and Options* (Rome: 2007), p. 53.

32. Ibid.

33. Danielle Nierenberg, *Happier Meals: Rethinking the Global Meat Industry*, Worldwatch Paper 171 (Washington, DC: Worldwatch Institute, 2005), p. 48.

34. CDC, "CDC Estimates of Foodborne Illness in the United States: CDC 2011 Estimates: Findings," at www.cdc.gov /foodborneburden/2011-foodborne-estimates.html, viewed 20 August 2012.

35. Ibid.

36. Ibid.

37. Grace et al., op. cit. note 30, citing ILRI, *Targeting Strategic Investment in Livestock Development as a Vehicle for Rural Livelihoods: Report to Bill and Melinda Gates Foundation* (Nairobi: 2009).

38. John O'Meara, "Alternative Cows Find their Dairy Niche," Rodale Institute, 2 May 2008, at www.rodaleinstitute.org/20080502/nf1.

Farm Animal Populations Continue to Grow (pages 53–56)

1. U.N. Food and Agriculture Organization (FAO), *FAOSTAT Statistical Database*, at www.faostat.fao.org, updated 23 February 2012.

2. Ibid.

3. M. W. Rosegrant et al., "Looking into the Future for Agriculture and AKST (Agricultural Knowledge, Science, and Technology)," in B. C. McIntyre et al., eds., *Agriculture at a Crossroads* (Washington, DC: Island Press, 2009), pp. 307–76.

4. FAO, *The State of Food and Agriculture 2009: Livestock in the Balance* (Rome: 2009), p. 9.

5. FAO, op. cit. note 1.

6. Ibid.

7. Ibid.

8. Ibid.

9. Ibid.

10. Ibid.

11. FAO, *Pro-Poor Livestock Policy Initiative: Livestock Sector Policies and Programmes in Developing Countries* (Rome: 2010) p. vii.

12. U.N. Population Division, *World Urbanization Prospects:*

The 2009 Revision Population Database, at www.esa.un.org /wup2009/unup, updated 2010.

13. Ibid.

14. World Bank, *World Bank Development Indicators*, at data .worldbank.org/indicator, updated 15 December 2011.

15. FAO, *Livestock's Long Shadow, Environmental Issues and Options* (Rome: 2007), p. 53.

16. Cees de Haan et al., "Structural Change in the Livestock Sector," in Henning Steinfeld et al., eds., *Livestock in a Changing Landscape* (Washington, DC: Island Press, 2010), p. 45; FAO, op. cit. note 1.

17. De Haan et al., op. cit. note 16, p. 45; FAO, op. cit. note 1.

18. U.S. Department of Agriculture, National Agricultural Statistics Service, at www.nass.usda.gov.

19. An Pan et al., "Red Meat Consumption and Mortality," *Archives of Internal Medicine,* published online 12 March 2012.

20. Humane Society of the United States, *The Impact of Industrialized Animal Agriculture on Rural Communities* (Washington, DC: 2009).

21. K. Winpisinger-Slay and R. Berry, *1998 Fly Investigation in Hardin, Marion, Wyandot, and Union Counties* (Columbus, OH: Ohio Department of Health, 1999).

22. "2011 Gulf of Mexico 'Dead Zone' Could Be Biggest Ever," *Science Daily*, 18 July 2011.

23. FAO, op. cit. note 4, p. 54.

24. Ibid., p. 5.

25. Mario Herrerro et al., "Improving Food Production from Livestock," in Worldwatch Institute, *State of the World 2011* (New York: W. W. Norton & Company, 2011), p. 157; FAO, op. cit. note 4, p. 56.

26. FAO, op. cit. note 15, p. 113.

27. FAO, op. cit. note 4, p. 84.

28. World Health Organization (WHO), "Zoonoses and Veterinary Public Health," at www.who.int/zoonoses/en, viewed 14 March 2012.

29. WHO, "Cumulative Number of Confirmed Human Cases of Avian Influenza A(H5N1) Reported to WHO," at www.who.int/influenza/human_animal_interface, viewed 14 March 2012.

30. WHO, "Outbreaks of E. Coli O104:H4 Infection: Update 30," at www.euro.who.int/en/what-we-do/health-topics, viewed 14 March 2012.

31. FAO, op. cit. note 4, p. 93.

32. Wallinga quoted in Danielle Nierenberg, *Happier Meals:*

Rethinking the Global Meat Industry, Worldwatch Paper 171 (Washington, DC: Worldwatch Institute, 2005), p. 48.

33. U.S. Environmental Protection Agency, "Dairy Production Systems," at www.epa.gov/agriculture/ag101, viewed 9 March 2012.

34. FAO, *The Status and Trends of Animal Genetic Resources 2010* (Rome: 2010).

35. FAO, Agriculture and Consumer Protection Department, "Spotlight/2006: Farm Animal Biodiversity," at www .fao.org/ag/magazine, September 2006.

36. Danielle Nierenberg, "The Keepers of Genetic Diversity: Meeting with Pastoralist Communities in Kenya" (blog), *Nourishing the Planet*, 10 November 2009.

37. FAO, op. cit. note 4, p. 15; FAO, op. cit. note 1.

38. FAO, Agriculture and Consumer Protection Department, "Spotlight/2006: Livestock Impacts on the Environment," at www.fao.org/ag/magazine, November 2006.

39. FAO, *Livestock Report 2006* (Rome: 2006), p. 13.

40. FAO, op. cit. note 15, p. xx.

41. Nierenberg, op. cit. note 32, p. 8.

42. R. Pelant et al., "Small Ruminants in Development: The Heifer Project International Experience in Asia," *Small Ruminant Research*, November 1999, pp. 249–57.

43. FAO, op. cit. note 4, p. 33; Jan de Wit et al., "Animal Manure: Asset or Liability?" *World Animal Review*, 1997, pp. 30–37.

44. International Energy Agency, "Access to Electricity," at www.iea.org/weo/electricity.asp, viewed 16 March 2012.

45. FAO, op. cit. note 11, pp. 15, 108, and 124; Herrerro et al., op. cit. note 25, p. 162.

Aquaculture Tries to Fill World's Insatiable Appetite for Seafood (pages 57–61)

1. U.N. Food and Agriculture Organization (FAO), *The State of the World Fisheries and Aquaculture 2012* (Rome: 2012). Data for 2011 are provisional estimates.

2. Ibid.

3. Ibid., p. 3.

4. Ibid.

5. Ibid.

6. Ibid.

7. Ibid.

8. Ibid.; FAO, *The State of the World Fisheries and Aquaculture 2008* (Rome: 2008), p. 176.

9. Rosamond L. Naylor, "Effect of Aquaculture on World Fish Supplies," *Nature*, 29 June 2000, p. 1,017.

10. FAO, op. cit. note 1, p. 3.

11. Ibid.

12. Ibid.

13. Ibid.

14. FAO, *Global Production Statistics 1950–2010*, updated 7 May 2012.

15. Ibid.

16. Ibid.

17. FAO, op. cit. note 1, p. 3.

18. Ibid.

19. Ibid.

20. Ibid.

21. Ibid.

22. Ibid., p. 84.

23. Ibid., p. 85.

24. Ibid.

25. Ibid., p. 84.

26. Ibid.

27. FAO, *2010 Yearbook of Fishery and Aquaculture Statistics* (Rome: 2012).

28. FAO, op. cit. note 14.

29. FAO, *Farming the Waters for People and Food, Proceedings of the Global Conference on Aquaculture 2010, Phuket, Thailand, 22–25 September 2010* (Rome: 2012), p. 5.

30. FAO, op. cit. note 1, pp. 41–42.

31. Ibid., p. 41.

32. Ibid.

33. Ibid., p. 46; FAO, *The State of the World Fisheries and Aquaculture 2010* (Rome: 2011), p. 29.

34. FAO, op. cit. note 1, p. 46.

35. Ibid.

36. John Vidal, "Senegal Revokes Licenses of Foreign Fishing Trawlers," (London) *Guardian*, 4 May 2012.

37. Ibid.

38. Ibid.

39. FAO, op. cit. note 1, p. 53.

40. Ibid.

41. Ibid.

42. Ibid.

43. Ibid.

44. Ibid.

45. Ibid.

46. Ibid., p. 59.

47. Ibid.

48. Ibid., p. 172.

49. Ibid.

50. Ibid., pp. 174, 177.

51. Ibid., p. 181; Matthias Halwart, "Aquaculture Topics and Activities: Aquaculture Feeds and Fertilizers," FAO Fisheries and Aquaculture Department, at www.fao.org /fishery/topic/13538/en, updated 27 May 2005.

52. FAO, op. cit. note 1, p. 192.

53. Ibid., p. 9.

54. Ibid.

55. Alexei Barrionuevo, "Chile's Antibiotics Use on Salmon Farms Dwarfs That of a Top Rival's," *New York Times*, 26 July 2009.

56. Oceana, "Chile Passes Legislative Reform on Salmon Escapes, Antibiotics," Aquaculture: Victories, at www.oceana .org/en/our-work/stop-ocean-pollution/aquaculture/victo ries, updated 2012.

57. FAO, op. cit. note 1, p. 30.

58. Ibid., p. 31.

59. Ibid.

60. Ibid.

61. Ibid., p. 32.

62. Sukhdev quoted in Sebastian Smith, "Oceans' Fish Could Disappear in 40 Years: UN," *Agence France Presse*, 17 May 2010.

Area Equipped for Irrigation at Record Levels, But Expansion Slows (pages 62–65)

1. U.N. Food and Agriculture Organization (FAO), *FAO-STAT Statistical Database*, at www.faostat.fao.org.

2. Ibid.

3. Calculated from S. Siebert et al., "Groundwater Use for Irrigation–A Global Inventory," *Hydrology and Earth System Sciences*, vol. 14 (2010), p. 1,868.

4. Ibid.

5. Ibid.

6. Ibid., p. 1863.

7. David Molden, ed., *Water for Food, Water for Life: A Comprehensive Assessment of Water Management for Agriculture* (London and Colombo, Sri Lanka: Earthscan and International Water Management Institute, 2007), p. 358.

8. FAO, *Climate Change, Water, and Food Security*, FAO Water Reports No. 36 (Rome: 2011), p. 16.

9. FAO, op. cit. note 1.

10. FAO, op. cit. note 8.

11. Ibid., p. 17.

12. Siebert et al., op. cit. note 3, p. 1863.

13. FAO, op. cit. note 8, p. 17.

14. World Bank, *Agriculture Investment Sourcebook: Agriculture and Rural Development* (Washington, DC: 2005), p. 370.

15. FAO, op. cit. note 8, p. 17.

16. Julia Apland Hitz, "The Worsening Water Crisis in Gujarat, India" (State of the Planet blog), Earth Institute, Columbia University, 18 January 2011.

17. FAO, op. cit. note 8, p. 17.

18. Sandra Postel, "Our Oversized Groundwater Footprint," Water Currents, *National Geographic*, August 2012.

19. Ibid.

20. Cynthia Podmore, "Irrigation Salinity—Causes and Impacts," State of New South Wales, Department of Industry and Investment, October 2009.

21. Ibid.

22. Ibid.

23. Ibid.

24. GRAIN, "Squeezing Africa Dry: Behind Every Land Grab is a Water Grab," June 2012.

25. Ibid.

26. Ibid.

27. Sandra Postel, "Drip Irrigation Expanding Worldwide," Water Currents, *National Geographic*, June 2012.

28. Ibid.

29. Ibid.

30. F. B. Reinders, "Micro-Irrigation: World Overview on Technology and Utilization," Keynote address at the 7th International Micro-Irrigation Congress in Kuala Lumpur, Malaysia, September 2006, p.1.

31. Postel, op. cit. note 27.

32. Reinders, op. cit. note 30, p. 7.

33. Postel, op. cit. note 27.

34. Ibid.

35. The World Food Prize, "Dr. Daniel Hillel Named 2012 World Food Prize Laureate," 12 June 2012, at www.world foodprize.org/en/press/news/index.cfm?action=display&newsID=18914.

36. Ibid.

37. Reinders, op. cit. note 30, p. 5.

38. Ibid., p. 6.

39. Postel, op. cit. note 27.

40. Ibid.

41. FAO, op. cit. note 8, p. 17.

42. Ibid., p. 18.

43. Ibid.

44. Ibid.

45. Ibid., p. 17.

46. Ibid.

Organic Agriculture Contributes to Sustainable Food Security (pages 66–68)

1. Helga Willer and Lukas Kilcher, eds., *The World of Organic Agriculture—Statistics and Emerging Trends 2012* (Bonn, Germany, and Frick, Switzerland: International Federation of Organic Agriculture Movements (IFOAM) and Research Institute of Organic Agriculture, 2012).

2. Ibid.

3. Ibid, p. 25.

4. Ibid.

5. Andrew Monk, "Sustainable and Resilient Social Structures for Change: The Organic Movement," *Social Alternatives*, vol. 30, no. 1 (2011), pp. 34–37; Willer and Kilcher, op. cit. note 1, p. 25.

6. IFOAM, "Definition of Organic Agriculture," at www .ifoam.org/growing_organic/definitions/doa/index.html.

7. Monk, op. cit. note 5, p. 34.

8. Kathleen Delate, *Fundamentals of Organic Agriculture* (Ames: Iowa State University, 2003), p. 2.

9. Ibid.

10. Rodale Institute, "The Farming Systems Trial: Celebrating 30 Years" (2011), at 66.147.244.123/~rodalein/wp-content/uploads/2012/12/FSTbookletFINAL.pdf.

11. Verena Seufert, Navin Ramankutty, and Jonathan Foley, "Comparing the Yields of Organic and Conventional Agriculture," *Nature*, 10 May 2012, pp. 229–32.

12. Christopher Barrett, "Measuring Food Insecurity," *Science*, 12 February 2012, pp. 825–28; Hettie Schönfeldt, N. Gibson, and H. Vermeulen, "The Possible Impact of Inflation on Nutritionally Vulnerable Households in a Developing Country using South Africa as a Case Study," *British Nutrition Foundation Nutrition Bulletin*, 20 August 2010, pp. 254–67.

13. The Government Office for Science, *The Future of Food and Farming: Challenges and Choices for Global Sustainability*, final report from the Foresight Global Food and Farming Futures project (London: 2011), p. 40.

14. Nadia Scialabba, *Organic Agriculture and Food Security*, International Conference on Organic Agriculture and Food Security (Rome: U.N. Food and Agriculture Organization [FAO], 2007).

15. Olivier De Schutter, "Agroecology and the Right to Food," report submitted to the Human Rights Council, 20 December 2010, p. 11.

16. Adrian Muller et al., *Reducing Global Warming and Adapting to Climate Change: The Potential of Organic Agriculture*, Working Paper No. 526 (Sweden: Göteburg University, 2012), pp. 1–7.

17. Janne Bengtsson, Johan Ahnström, and Ann-Christin Weibull, "The Effects of Organic Agriculture on Biodiversity and Abundance: A Metaanalysis," *Journal of Applied Ecology* (April 2005), pp. 261–69.

18. Willer and Kilcher, op. cit. note 1, p. 28.

19. Ibid.

20. Catherine Greene, Edward Slattery, and William McBride, "America's Organic Farmers Face Issues and Opportunities," *Amber Waves*, June 2010.

21. Organic Trade Association, "Consumer-Driven U.S. Organic Market Surpasses $31 Billion in 2011," press release (Washington, DC: 23 April 2012).

22. Carl Haub and James Gribble, "Population Bulletin: The World at 7 Billion," *Population Reference Bureau*, July 2011, pp. 1–12.

23. Willer and Kilcher, op. cit. note 1, p. 26.

24. Ibid., Executive Summary.

25. Ibid., p. 27.

26. International Assessment of Agricultural Knowledge, Science and Technology Development (IAASTD), *Agriculture at a Crossroads: Sub-Saharan Africa (SSA) Report* (Washington, DC: Island Press, 2009).

27. Dennis Garrity et al., "Evergreen Agriculture: A Robust Approach to Sustainable Food Security in Africa," *Food Security*, September 2010, pp. 197–214.

28. Leo Stroosnijder, "Modifying Land Management in Order to Improve Efficiency of Rain Water Use in the African Highlands," *Soil and Tillage Research*, May 2009, pp. 247–56.

29. Sue Edwards et al., "Successes and Challenges in Ecological Agriculture: Experiences from Tigray, Ethiopia," in Lim Li Ching et al., eds., *Climate Change and Food Systems Resilience in Sub-Saharan Africa* (Rome: FAO, 2011), pp. 1–58.

30. Willer and Kilcher, op. cit. note 1, p. 28.

31. Ibid.

32. Jitendra Pandey and Ashima Singh, "Opportunities and Constraints in Organic Farming: An Indian Perspective," *Journal of Scientific Research* (Baranas Hindu University), vol. 56 (2012), pp. 47–72.

33. Alice Beban, *Organic Agriculture and Farmer Wellbeing: A Case Study of Cambodian Small-scale Farmers* (Palmerston North, New Zealand: Massey University, Institute of Development Studies, 2009), p. 2.

34. Ibid., p. 7.

35. The Government Office for Science, op. cit. note 13, p. 9.

36. IAASTD, op. cit. note 26, p. 2.

37. Pandey and Singh, op. cit. note 32, p. 52.

Investing in Women Farmers (pages 69–73)

1. U.N. Food and Agriculture Organization (FAO), *2010–2011 The State of Food and Agriculture—Women in Agriculture: Closing the Gender Gap for Development* (Rome: 2011).

2. World Bank, *World Development Indicators & Global Development Finance*, at data.worldbank.org.

3. FAO, op. cit. note 1.

4. Kathleen Masterson, "U.S. Women Farmers," *All Things Considered*, National Public Radio, 30 March 2011.

5. International Labour Organization (ILO), *Empower Rural Women—End Poverty and Hunger*, background paper (Geneva: 2012), pp. 1–4.

6. FAO, op. cit. note 1.

7. United Nations Girl's Education Institute (UNGEI), *Engendering Empowerment: Education and Equality* (New York: UNICEF, 2012), p. 1.

8. FAO, op. cit. note 1.

9. FAO, *The Role of Women in Crop, Livestock Fisheries, and Agroforestry*, synthesis report, at fao.org/docrep/x0176e/x0176e05.htm; Salma Ahmed and Pushkar Maitra, "Gender Wage Discrimination in Rural and Urban Labour Markets of Bangladesh," *Oxford Development Studies*, March 2010, pp. 83–112.

10. Beatrice Costa, *Her Mile—Women's Rights and Access to Land: The Last Stretch of Road to Eradicate Hunger* (Rome: Action Aid, 2011), pp. 4–7.

11. Mark Curtis, *Fertile Ground: How Governments and Donors Can Halve Hunger by Supporting Small Farmers* (London: Action Aid UK, 2010), p. 16.

12. Ibid., p. 17.

13. Economist Intelligence Unit, *Food Security Index*, at foodsecurityindex.eiu.com/Index, viewed 10 July 2012.

14. Economist Intelligence Unit, "New Study Spotlights Opportunities and Barriers for Working Women Worldwide," at eiu.com/site_info.asp?info_name=womens_economic_opportunity&page=noads.

15. United Nations, "Women at a Glance," at www.un.org/ecosocdev/geninfo/women/women96.htm.

16. ILO, op. cit. note 5.

17. Ibid.

18. World Health Organization, "Adolescent Pregnancy," fact sheet, at who.int/mediacentre/factsheets/fs364/en.

19. UNICEF, "UNICEF Says Education for Women and Girls a Lifeline to Development," press release (New York: 4 May 2011).

20. Nazmul Chaudhury and Dilip Parajuli, "Conditional Cash Transfers and Female Schooling: The Impact of the Female School Stipend Programme on Public School Enrolments in Punjab, Pakistan," *Applied Economics*, vol. 42, no. 28 (2010), pp. 3,565–83.

21. Priya Bhagowalia et al., "Unpacking the Links between Women's Empowerment and Child Nutrition: Evidence Using Nationally Representative Data from Bangladesh," presented at the Agricultural and Applied Economics Association Annual Meeting, Denver, CO, 25–27 July 2010.

22. Babatunde Omilola, *Patterns and Trends of Child and Maternal Nutrition Inequalities in Nigeria*, Discussion Paper 00968 (Washington, DC: International Food Policy Research Institute [IFPRI], 2010), p. 36.

23. Agnes Reynes Quisumbing and Lauren Pandolfelli, "Promising Approaches to Address the Needs of Poor Female Farmers: Resources, Constraints, and Interventions," *World Development*, vol. 38, no. 4 (2010), pp. 581–92.

24. FAO, op. cit. note 1, pp. 34–35.

25. F. M. Kinkingninhoun-Mêdagbé et al., "Gender Discrimination and Its Impact on Income, Productivity, and Technical Efficiency: Evidence from Benin," *Agriculture and Human Values*, vol. 27, no.1 (2010), pp. 57–69.

26. FAO, op. cit. note 1, p. 23.

27. Ibid., p. 26.

28. FAO, "Men and Women in Agriculture: Closing the Gap: Key Facts," at www.fao.org/sofa/gender/key-facts/en, viewed 5 July 2012.

29. Hema Swaminathan, Rahul Lahoti, and Suchitra J. Y., *Women's Property, Mobility, and Decisionmaking: Evidence from Rural Karnataka, India*, Discussion Paper no. 01188 (Washington, DC: IFPRI, 2012), p. 7.

30. OECD Development Center, *2012 SIGI Social Institutions and Gender Index: Understanding the Drivers of Gender Inequality* (Paris: Organisation for Economic Co-operation and Development, 2012).

31. Economist Intelligence Unit, op. cit. note 13.

32. Calvert Foundation, "Women Investing in Women Initiative," at www.calvertfoundation.org/the-economic-power-of-women.

33. Kirrin Gill et al., *Invisible Market: Energy and Agricultural Technologies for Women's Economic Advancement* (Washington, DC: International Center for Research on Women (ICRW), 2012), p. 5.

34. Ibid., p. 21.

35. FAO, op. cit. note 1.

36. Ibid.

37. Ibid.

38. Krista Jacobs et al., *Gender Differences in Asset Rights in KwaZulu-Natal, South Africa* (Washington, DC: ICRW, 2011).

39. FAO, *Gender and Land Rights: Understanding Complexities; Adjusting Policies*, Policy Brief (Rome: 2010).

40. FAO, op. cit. note 1, p. 47.

41. Ibid., p. 48.

42. Ricardo Hausmann, Laura D. Tyson, and Saadia Zahidi, *The Global Gender Gap Report 2011* (Geneva: World Economic Forum, 2011).

43. Costa, op. cit. note 10.

Foreign Investment in Agricultural Land Down from 2009 Peak (pages 74–77)

1. Land Matrix Project, *Land Matrix Project Database*, at www.landportal.info/landmatrix, updated April 2012.

2. Agricultural land from U.N. Food and Agriculture Organization (FAO), *FAOSTAT Resources*, at www.fao.org/economic/ess/ess-publications/ess-yearbook/ess-yearbook2010/yearbook2010-reources/en/, updated 2010.

3. Land Matrix Project, op. cit. note 1.

4. Ibid.

5. Ibid.

6. GRAIN, *Grain Land Grab Deals* (Barcelona: 2012).

7. Land Matrix Project, op. cit. note 1.

8. Ward Anseeuw et al., *Transnational Land Deals for Agriculture in the Global South: Analytical Report Based on the Land Matrix Database* (Bern, Germany; Montpellier, France; and Hamburg, Germany: CDE/CIRAD/GIGA, 2012).

9. Land Matrix Project, op. cit. note 1.

10. Lorenzo Cotula et al., *Land Grab or Development Opportunity? Agricultural Investment and International Land Deals in Africa* (Rome: FAO, International Institute for Environment and Development, and International Fund for Agricultural Development, 2009); U.N. Conference on Trade and Development, *Inward and Outward Foreign Direct Investment Flows, Annual, 1970–2010,* at unctadstat.uctad.org.

11. Land Matrix Project, op. cit. note 1.

12. Ibid.

13. Ibid.

14. Ibid.

15. Ibid.

16. Ibid.

17. Anseeuw et al., op. cit. note 8.

18. Ibid.

19. Ibid.

20. Ibid.

21. Ibid.

22. Ibid.

23. Ibid.

24. Land Matrix Project, op. cit. note 1.

25. Calculated from GRAIN, op. cit. note 6.

26. Land Matrix Project, op. cit. note 1.

27. Ibid.

28. Ward Anseeuw et al., *Land Rights and the Rush for Land: Findings of the Global Commercial Pressures on Land Research Project* (Rome: International Land Coalition, 2012); FAO, *FAO Food Price Index,* at faostat.fao.org, updated April 2012.

29. Anseeuw et al., op. cit. note 8.

30. Ibid.

31. Anseeuw et al., op. cit. note 28.

32. FAO, op. cit. note 28.

33. Anseeuw et al., op. cit. note 8.

34. Lorenzo Cotula and Sonja Vermeulen, "Deal or No Deal: The Outlook for Agricultural Land Investment for Africa," *International Affairs*, November 2009, pp. 1,233–47.

35. Anseeuw et al., op. cit. note 28.

36. Klaus Deininger et al., *Rising Global Interest in Farmland: Can It Yield Sustainable and Equitable Benefits?* (Washington, DC: World Bank, 2010).

37. Oxfam International, *Land and Power. The Growing Scandal Surrounding the New Wave of Investments in Land* (Oxford: Oxfam International, 2011).

38. Timothy A. Wise and Sophia Murphy, *Resolving the Food Crisis: Assessing Global Policy Reforms Since 2007* (Medford, MA: Institute for Agriculture and Trade Policy and Global Development and Environment Institute, 2012).

Wage Gap Widens as Wages Fail to Keep Pace with Productivity (pages 80–83)

1. International Labour Organization (ILO), *Global Employment Trends 2013: Recovering from a Second Jobs Dip* (Geneva: 2013), p. 166.

2. Ibid.

3. Ibid., pp. 174, 188.

4. ILO, *Global Wage Report 2012/13: Wages and Equitable Growth* (Geneva: 2013), p. xiii.

5. Ibid.

6. Ibid., p. xiii.

7. Ibid.

8. Ibid..

9. Ibid., Preface.

10. Ibid., pp. xiii-xiv.

11. U.S. Bureau of Labor Statistics (BLS), "International Comparisons of Hourly Compensation Cost in Manufacturing, 2011," 19 December 2012, at www.bls.gov/news.release/pdf/ichcc.pdf.

12. Ibid.

13. Ibid.

14. Ibid.

15. Ibid.

16. Ibid.

17. Ibid.

18. Ibid.

19. ILO, op. cit. note 4, Preface.

20. Ibid., p. xiv.

21. Figure 3 from ILO, "The Widening Gap between Wages and Labour Productivity," 7 December 2012, at www.ilo.org /global/research/global-reports/global-wage-report/2012 /charts/WCMS_193307/lang--en/index.htm.

22. ILO, op. cit. note 4, p. 46.

23. Wolfgang Lieb, "'Nie gab es mehr Erwerbstätige'—Propaganda mit Zahlen," *Nachdenkseiten*, 4 January 2013.

24. Deutscher Gewerkschaftsbund, "Leiharbeit: Was heißt hier schon 'vorübergehend'?" 26 October 2012.

25. ILO, op. cit. note 4, p. xv.

26. Ibid., p. 44.

27. Lawrence Mishel and Natalie Sabadish, *CEO Pay and the Top 1%: How Executive Compensation and Financial-Sector Pay Have Fueled Income Inequality*, EPI Issue Brief No. 331 (Washington, DC: Economic Policy Institute, 2 May 2012), p. 1.

28. Ibid., p. 1.

29. Ibid., p. 2.

30. Ibid.

31. Till van Treek, *Did Inequality Cause the U.S. Financial Crisis?* IMK Working Paper No. 91 (Düsseldorf, Germany: April 2012); ILO, *World of Work Report 2011: Making Markets Work for Jobs*, Summary, preprint edition (Geneva: 31 October 2011), p. 3.

32. Till van Treeck and Simon Sturn, *Income Inequality as a Cause of the Great Recession? A Survey of Current Debates*, Conditions of Work and Employment Series No. 39 (Geneva: ILO, 2012), p. 90.

33. Ibid.

34. Ibid., p. 46; ILO, *Global Employment Trends 2012: Preventing a Deeper Jobs Crisis* (Geneva: 2012), p. 46.

35. ILO, op. cit. note 4, p. xiv.

36. ILO, *Global Wage Report 2010/11: Wages Policies in Times of Crisis*, Executive Summary (Geneva: 2010), p. 4.

37. Ibid.

Metals Production Recovers (pages 84–87)

1. U.S. Geological Survey (USGS), *Mineral Commodity Summaries* (Reston, VA: 2012).

2. Ibid.

3. R. B. Gordon, M. Bertram, and T. E. Graedel, "Metals Stocks and Sustainability," *Proceedings of the National Academy of Sciences*, 31 January 2006, pp. 1,209–14.

4. Ibid.

5. World Steel Association, *Steel Statistical Yearbook 2011* (Brussels: 2011).

6. Ibid.

7. Ibid.

8. Worldwatch analysis based on data in USGS, op. cit. note 1.

9. World Steel Association, *Steel Statistical Yearbooks* (Brussels: various years).

10. W. D. Menzie et al., *The Global Flow of Aluminum from 2006 through 2025* (Reston, VA: USGS, 2010).

11. T. E. Graedel, *Metal Stocks in Society: A Scientific Synthesis* (Paris: International Resource Panel, United Nations Environment Programme (UNEP), 2010).

12. World Steel Association, op. cit. note 9.

13. Menzie et al., op. cit. note 10.

14. Ibid.

15. Ibid.

16. Gordon, Bertram, and Graedel, op. cit. note 3.

17. Ibid.

18. Ibid.

19. Thomas Graedel et al., *Recycling Rates of Metals: A Status Report*, A Report of the Working Group on Global Metal Flows (Paris: International Resource Panel, UNEP, 2011).

20. Ibid.

21. Ibid.

22. Ibid.

23. USGS, "Metal Stocks in Use in the United States," Fact Sheet 2005-3090 (Reston, VA: 2005).

24. Ibid.

25. Ibid.

26. Ibid.

27. Menzie et al., op. cit. note 10.

28. Ibid.

29. Yasuhiko Hotta, "Sound Material Cycle Society from Japan to Asia," presented at International Green Technology and Purchasing Conference, Kuala Lumpur, Malaysia, 15–16 October 2010.

Municipal Solid Waste Growing (pages 88–90)

1. Daniel Hoornweg and Perinaz Bhada-Tata, *What a Waste: A Global Review of Solid Waste Management* (Washington, DC: World Bank, 2012).

2. Ibid.

3. Ibid.

4. Ibid.

5. Ibid.

6. U.N. Environment Programme (UNEP), *Towards a Green Economy: Pathways to Sustainable Development and Poverty Eradication* (Paris: 2011), p. 294.

7. Hoornweg and Bhada-Tata, op. cit. note 1.

8. Ibid.

9. In Table 1, United States is a Worldwatch conversion to metric tons from U.S. Environmental Protection Agency (EPA), "Municipal Solid Waste Generation, Recycling, and Disposal in the United States: Facts and Figures for 2010," at www.epa.gov/osw/nonhaz/municipal/pubs/msw_2010 _rev_factsheet.pdf; other countries from Hoornweg and Bhada-Tata, op. cit. note 1.

10. Worldwatch analysis based on data in Hoornweg and Bhada-Tata, op. cit. note 1.

11. Ibid.

12. Ibid.

13. Ibid.

14. Ibid.

15. Ibid.

16. UNEP, op. cit. note 6.

17. EPA, op. cit. note 9.

18. In Table 2, United States from EPA, op. cit. note 9; other countries from Hoornweg and Bhada-Tata, op. cit. note 1.

19. Hoornweg and Bhada-Tata, op. cit. note 1.

20. UNEP, op. cit. note 6, p. 303.

21. Hoornweg and Bhada-Tata, op. cit. note 1.

22. UNEP, op. cit. note 6.

23. Ibid.

24. EPA, op. cit. note 9.

25. Energy from Hoornweg and Bhada-Tata, op. cit. note 1; water and emissions from UNEP, op. cit. note 6.

26. Trees and water from UNEP, op. cit. note 6; gasoline from EPA, op. cit. note 9.

27. Yasuhiko Hotta, "Sound Material Cycle Society from Japan to Asia," presented at International Green Technology and Purchasing Conference, Kuala Lumpur, Malaysia, 15–16 October 2010.

28. Ibid.

Losses from Natural Disasters Reach New Peak in 2011 (pages 91–95)

1. Munich Re calculation, based on NatCatSERVICE database, 2012.

2. Ibid.

3. Munich Re, *Topics Geo: Natural Catastrophes 2011— Analyses, Assessments, Positions* (Munich: 2012), pp. 50–51.

4. Munich Re, op. cit. note 1.

5. Munich Re, op. cit. note 3.

6. Ibid.

7. Munich Re, op. cit. note 1.

8. Ibid.

9. Ibid.

10. Ibid.

11. Munich Re, op. cit. note 3.

12. Munich Re, op. cit. note 1.

13. Ibid.

14. Ibid.

15. Munich Re, op. cit. note 3.

16. Ibid.

17. Munich Re, op. cit. note 1.

18. Munich Re, op. cit. note 3.

19. Munich Re, op. cit. note 1.

20. Munich Re, op. cit. note 3.

21. Munich Re, op. cit. note 1.

22. Munich Re, op. cit. note 3, p. 32.

23. Ibid.

24. Ibid., p. 43.

25. Munich Re, op. cit. note 1.

26. Munich Re, op. cit. note 3.

27. Munich Re, op. cit. note 1.

28. Ibid.

29. Ibid.

30. Ibid.

31. Ibid.

32. Ibid.

33. Munich Re, op. cit. note 3, p. 40.

34. Ibid. , p. 42.

35. Ibid.

36. Munich Re, op. cit. note 1.

37. Ibid.

38. U.S. Geological Survey (USGS), "Significant Earthquakes of the World 2011," at earthquake.usgs.gov/earthquakes.

39. Munich Re, op. cit. note 1.

40. Ibid.

41. Munich Re, op. cit. note 3, p. 42.

42. Ibid., p. 27.

43. Munich Re, op. cit. note 1.

44. Munich Re, op. cit. note 3, p. 26.

45. Munich Re, op cit. note 1.

46. Ibid.

47. USGS, op. cit. note 38; Erol Kalkan, "The Christchurch Earthquake's Hidden Secrets," *Natural Hazards Observer*, January 2012, pp. 1, 12–14.

48. Munich Re, op. cit. note 1.

49. Ibid.

50. Ibid.

51. Ibid.

52. Ibid.

53. Ibid.

54. Ibid.

55. Ibid.

56. Ibid.

57. Ibid.

The Looming Threat of Water Scarcity (pages 96–100)

1. UN Water, "Water for Life 2005-2015: Water Scarcity," at www.un.org/waterforlifedecade/scarcity.shtml, viewed 20 February 2013.

2. Ibid.

3. U.N. Food and Agriculture Organization (FAO), *Coping with Water Scarcity: An Action Framework for Agriculture and Food Security,* FAO Water Report 38 (Rome: 2012).

4. Ibid.

5. UN Water, op. cit. note 1.

6. Ibid.

7. World Water Assessment Programme (WWAP), *World Water Development Report, Vol. 1: Managing Water under Uncertainty and Risk* (Paris: UNESCO, 2012).

8. Ibid.

9. Ibid.

10. Ibid.

11. Ibid.

12. FAO, *AQUASTAT*, at www.fao.org/nr/water/, viewed 1 March 2013.

13. WWAP, op. cit. note 7.

14. Worldwatch calculation based on total renewable water resources from FAO, op. cit. note 12; total renewable water resources from European Environment Agency, "Water Availability," at www.eea.europa.eu/themes/water/water-resources/water-availability, 18 February 2008; population data from U.N. Department of Economic and Social Affairs, at esa.un.org/unpd/wpp/Excel-Data/population.htm, viewed 1 March 2013.

15. WWAP, op. cit. note 7.

16. UN Water, "Statistics: Graphs & Maps," at www.unwater.org/statistics_use.html, viewed on 22 February 2013.

17. WWAP, op. cit. note 7.

18. FAO, *AQUASTAT*, at www.fao.org/nr/water/aquastat/water_use/index.stm, viewed 1 March 2013.

19. FAO, op. cit. note 3.

20. WWAP, op. cit. note 7.

21. Ibid.

22. S. Siebert et al., "Groundwater Use for Irrigation—A Global Inventory," *Hydrology and Earth System Sciences*, vol. 14 (2010), pp. 1,863–80.

23. A. Jägerskog and T. Jønch Clausen, eds., *Feeding a Thirsty World: Challenges and Opportunities for a Water and Food Secure World* (Stockholm: Stockholm International Water Institute, 2012).

24. Ibid.

25. UNICEF, "Water in India: Situation and Prospects," at www.unicef.org/india/media_8098.htm, viewed 1 March 2013.

26. Jägerskog and Jønch Clausen, op. cit. note 23.

27. System of Rice Intensification, "Frequently Asked Questions," at sri.ciifad.cornell.edu/aboutsri/FAQs1.html#what arethemain, viewed 6 March 2013.

28. Ibid.

29. WWAP, op. cit. note 7; Pacific Institute, "Water Content of Things," at www.worldwater.org/data20082009/Table19 .pdf, viewed 3 March 2013.

30. FAO, op. cit. note 3.

31. WWAP, op. cit. note 7.

32. Ibid. Percentage calculated by the WWAP report based on World Energy Council numbers.

33. Ibid.

34. Hugh Turral, Jacob Burke, and Jean-Marc Faurès, *Climate Change, Water and Food Security* (Rome: FAO, 2011).

35. WWAP, op. cit. note 7.

Advertising Spending Continues Gradual Rebound, Driven by Growth in Internet Media (pages 101–03)

1. Zenith Optimedia, "ZenithOptimedia Forecasts 4.1% Growth in Global Adspend in 2013," press release (3 December 2012); Figure 1 compiled from ZenithOptimedia advertising expenditures forecasts from 2007 to 2013 converted to 2011 dollars where applicable using Bureau of Labor Statistics CPI Inflation Calculator.

2. Zenith Optimedia, op. cit. note 1.

3. Ibid.

4. Ibid.

5. Ibid.

6. Ibid.

7. Ibid.

8. PQ Media, *PQ Media U.S. Mobile & Social Media Forecast 2012–16* (Stamford, CT: 2012); Zenith Optimedia, op. cit. note 1.

9. Zenith Optimedia, op. cit. note 1.

10. Ibid.

11. E. Wright et al., "The Lasting Effects of Social Media Trends on Advertising," *Journal of Business & Economics Research*, November 2010; J. Colliander and M. Dahlén, "Following the Fashionable Friend: The Power of Social Media—Weighing Publicity Effectiveness of Blogs versus Online Magazines," *Journal of Advertising Research*, March 2011, pp. 313–20.

12. A. Molnar et al., "Schools Inundated in a Marketing-Saturated World," in J. A. Sandlin and P. McLaren, eds., *Critical Pedagogies of Consumption* (New York: Routledge,

2010), p. 83.

13. PQ Media, *PQ Media Global Product Placement Spending Forecast 2012–2016* (Stamford, CT: 2012).

14. Ibid.

15. N. Zmuda, "Ad Spending Rises 3% in 2012: Kantar Media Report," *Advertising Age*, 11 March 2013.

16. Ibid.

17. J. A. Sandlin and P. McLaren, "Introduction: Exploring Consumption's Pedagogy and Envisioning a Critical Pedagogy of Consumption—Living and Learning in the Shadow of the 'Shopocalypse'," in Sandlin and McLaren, op. cit. note 12, p. 8; J. A. Hill, "Endangered Childhoods: How Consumerism is Impacting Child and Youth Identity," *Media Culture & Society*, April 2011, pp. 347–62.

18. Hill, op. cit. note 17.

19. R. Farahmandpur, "Teaching Against Consumer Capitalism in the Age of Commercialization and Corporatization of Public Education," in Sandlin and McLaren, op. cit. note 12, p. 64.

20. Campaign for a Commercial-Free Childhood, "2013 School Bus Ad Action Center," at www.commercialfree childhood.org/action/2013schoolbusads, viewed 28 February 2013; Public Citizen's Commercial Alert, "School Buses," at www.commercialalert.org/issues/education/school-buses, viewed 28 February 2013.

21. Campaign for a Commercial-Free Childhood, op. cit. note 20; Public Citizen's Commercial Alert, op. cit. note 20; Farahmandpur, op. cit. note 19.

22. Institute of Medicine, *Food Marketing to Children and Youth: Threat or Opportunity?* (Washington, DC: National Academies Press, 2006), p. 8.

23. L. Pappas-Taffer and A. Miller, "Direct-to-Consumer Advertising of Prescription Medications: Misguided 'Autonomy' in the Information Age," *Dermatoethics*, 2012, pp. 13–17; B. Mintzes, "Advertising of Prescription-Only Medicines to the Public: Does Evidence of Benefit Counterbalance Harm?" *Annual Review of Public Health*, April 2012, pp. 259–77.

24. C. L. Ventola, "Direct-to-Consumer Pharmaceutical Advertising: Therapeutic or Toxic?" *Pharmacy and Therapeutics*, October 2011, pp. 669–674, 681–84.

25. A. B. Seidenberg et al., "Storefront Cigarette Advertising Differs by Community Demographic Profile," *American Journal of Health Promotion*, July/August 2012, pp. e26–e31.

26. Ibid.

27. R. T. Wilson and B. D. Till, "Targeting of Outdoor Alcohol Advertising: A Study Across Ethnic and Income

Groups," *Journal of Current Issues & Research in Advertising*, vol. 33, no. 2 (2012), pp. 267–81.

28. W. S. Tsai and M. Shumow, "Representing Fatherhood and Male Domesticity in American Advertising," *Interdisciplinary Journal of Research in Business*, August 2011, p. 38.

29. H. Paek, M. R. Nelson, and A. M. Vilela, "Examination of Gender-role Portrayals in Television Advertising across Seven Countries," *Sex Roles*, February 2011, pp. 192–207.

30. Ibid.

31. J. Neff, "Green-Marketing Revolution Defies Economic Downturn," *Advertising Age*, 20 April 2009.

32. S. Clifford and A. Martin, "As Consumers Cut Spending, 'Green' Products Lose Allure," *New York Times*, 21 April 2011.

33. Federal Trade Commission, "FTC Issues Revised 'Green Guides'," press release (Washington, DC: 1 October 2012).

34. Ibid.

Emerging Co-operatives (pages 106–08)

1. International Co-operative Alliance (ICA), *Global 300 Report 2010: The World's Major Co-operatives and Mutual Businesses* (Geneva: 2010).

2. Ibid.

3. Ibid.

4. Kimberly A. Zeuli and Robert Cropp, *Co-operatives: Principles and Practices in the 21st Century* (Madison, WI: University of Wisconsin, 2004).

5. University of Wisconsin Center for Co-operatives, "Co-operatives in the U.S. Economy," section of "Research on the Economic Impact of Co-operatives," study described at reic.uwcc.wisc.edu/issues, viewed 8 February 2012.

6. ICA, *Volume of International Cooperation, Vol. 100, No. 1* (Geneva: 2007).

7. Ibid.

8. Ed Mayo, *Global Business Ownership 2012: Members and Shareholders Across the World* (Manchester, U.K.: Co-operatives UK, 2012).

9. Ibid.

10. Ibid.

11. Ibid.

12. ICA, op. cit. note 1.

13. Johnston Birchall and Lou Hammond Ketilson, *Resilience of the Cooperative Business Model in Times of Crisis* (Geneva: International Labour Organization (ILO), Sustainable

Enterprise Programme, 2009).

14. Ibid.

15. Ibid.

16. ICA, op. cit. note 6.

17. Mayo, op. cit. note 8. Figure 1 from ibid. and from ICA, op. cit. note 1.

18. Nataliya Mylenko et al., *Financial Access 2010: the State of Financial Inclusion Through the Crisis* (Washington, DC: World Bank Group, 2010).

19. Ibid.

20. Benedicte Fonteneau et al., *THE READER 2011: Social and Solidarity Economy: Our Common Road towards Decent Work*, in support of the second edition of the Social and Solidarity Economy Academy, 24–28 October 2011, Montreal, Canada (Turin: ITC ILO, 2011).

21. World Council of Credit Unions, at www.woccu.org/publications/statreport.

22. Ibid.

23. Ibid.

24. Mylenko et al., op. cit. note 18.

25. Ibid.

26. Birchall and Hammond Ketilson, op. cit. note 13.

27. Ibid.

28. International Year of Cooperatives 2012, at social.un.org/coopsyear.

29. National Cooperative Business Association, "Support the National Cooperative Development Act," at www.ncba.coop.

Climate Change Migration Often Short-Distance and Circular (pages 109–12)

1. Kari Lyderson, "Scientists: Pace of Climate Change Exceeds Estimates," *Washington Post*, 15 February 2009.

2. "Convention relating to the Status of Refugees" and "Protocol relating to the Status of Refugees," at www.untreaty.un.org/cod/avl/ha/prsr/prsr.html.

3. U.N. Development Programme, *Human Development Report 2009* (New York: 2009).

4. Ibid.

5. U.N. High Commissioner for Refugees (UNHCR), "Populations of Concern to UNHCR," at www.unhcr.org/4ec230f516.html.

6. UNHCR, "Internally Displaced People," at www.unhcr.org/pages/49c3646c146.html.

7. Christian Aid, *Human Tide: The Real Migration Crisis* (London: 2007).

8. Richard Black et al., "The Effect of Environmental Change on Human Migration," *Global Environmental Change*, December 2011, pp. S3–S11.

9. Richard Black et al., "Climate Change: Migration as Adaptation," *Nature*, 27 October 2011, pp. 447–49.

10. Robert Engelman, *MORE: Population, Nature, and What Women Want* (Washington, DC: Island Press, 2008).

11. "Introduction: Migration and Climate Change," in Etienne Piguet, Antoine Pécoud, and Paul de Guchteneire, eds., *Migration and Climate Change* (Cambridge, U.K.: UNESCO and Cambridge University Press, 2011).

12. Robert A. McLeman and Lori M. Hunter, "Migration in the Context of Vulnerability and Adaptation to Climate Change: Insights from Analogues," *Wiley Interdisciplinary Reviews: Climate Change*, vol. 1, no. 3 (2010), pp. 450–61.

13. Katha Kartiki, "Climate Change and Migration: A Case Study from Rural Bangladesh," *Gender & Development*, March 2011, pp. 23–38.

14. Ibid.

15. Clark L. Gray and Valerie Mueller, "Natural Disasters and Population Mobility in Bangladesh," *Proceedings of the National Academy of Sciences*, 2 April 2012, p. 6,004.

16. Kartiki, op. cit. note 13.

17. Gray and Mueller, op. cit. note 15.

18. Kartiki, op. cit. note 13.

19. Shareen Joshi and T. Paul Schultz, *Family Planning as an Investment in Development: Evaluation of a Program's Consequences in Matlab, Bangladesh* (New Haven, CT: Institute for the Study of Labor and Yale University–Economic Growth Center, February 2007).

20. Colette Mortreux and Jon Barnett, "Climate Change, Migration and Adaptation in Funafuti, Tuvalu," *Global Environmental Change*, vol. 19 (2009), pp. 105–12.

21. Haakon Lein, "Hazards and 'Forced' Migration in Bangladesh," *Norwegian Journal of Geography*, vol. 54 (2000), p. 127.

22. Koko Warner et al., *In Search of Shelter: Mapping the Effects of Climate Change on Human Migration and Displacement* (New York: UN University, CARE International, Columbia University, UNHCR, and World Bank, 2009).

23. Ibid.

24. Ibid.

25. Deborah Balk et al., "Mapping Urban Settlements and the Risks of Climate Change in Africa, Asia and South America," in José Miguel Guzmán et al., eds., *Population Dynamics and Climate Change* (New York: United Nations Population Fund, 2009), pp. 80–97.

26. Ibid.

27. Gordon McGranahan, Deborah Balk, and B. Anderson, *Low Elevation Coastal Zone (LECZ) Urban-Rural Population Estimates, Global Rural-Urban Mapping Project (GRUMP), Alpha Version* (Palisades, NY: NASA Socioeconomic Data and Applications Center, 2007).

28. Asian Development Bank, *Key Indicators for Asia and the Pacific 2012* (Mandaluyong City, Philippines: 2012).

29. Balk et al., op. cit. note 25.

30. Ibid.

31. Katherine J. Curtis and Annemarie Schneider, "Understanding the Demographic Implications of Climate Change: Estimates of Localized Population Projections under Future Scenarios of Sea-Level Rise," *Population & Environment*, September 2011, pp. 28–54.

32. Ibid.

Urbanizing the Developing World (pages 113–16)

1. U.N. Department of Economic and Social Affairs (UN-DESA), *Population Distribution, Urbanization, Internal Migration and Development: An International Perspective* (New York: 2011).

2. Ibid.

3. UN-HABITAT, *State of the World's Cities 2008/2009* (Nairobi: 2008).

4. UN-DESA, *World Urbanization Prospects 2011* (New York: 2012).

5. Ibid.

6. Ibid.

7. Ibid.

8. Ibid.

9. UN-HABITAT, *Sustainable Urbanization* (Nairobi: 2012).

10. Ibid.

11. UN-DESA, op. cit. note 4.

12. U.N. Population Fund (UNFPA), *State of World Population 2011* (New York: 2011).

13. UN-DESA, op. cit. note 4.

14. Population Reference Bureau, *2012 World Population Data Sheet* (Washington, DC: 2012).

15. UN-DESA, op. cit. note 4.

16. Ibid.

17. Asian Development Bank, *Green Urbanization in Asia* (Manila: 2012).

18. Ibid.

19. UN-DESA, op. cit. note 4.

20. Asian Development Bank, op. cit. note 17.

21. Robert Malone and Tom Van Riper, "The World's Densest Cities," *Forbes*, 14 December 2007.

22. David Dodman, "Urban Density and Climate Change," prepared for UNFPA, 2 April 2009.

23. Elizabeth Eaves, "Two Billion Slum Dwellers," *Forbes*, 11 June 2007.

24. UN-HABITAT, *State of the World's Cities 2010/2011* (Nairobi: 2010).

25. Ibid.

26. World Health Organization (WHO), *Hidden Cities: Unmasking and Overcoming Health Inequities in Urban Settings* (Geneva: 2010).

27. Ibid.

28. Cities Alliance, "World Statistics Day: A Look at Urbanisation," October 2010, at www.citiesalliance.org.

29. WHO, op. cit. note 26.

30. Eaves, op. cit. note 23.

31. Ibid.

32. Charles Kenny, "In Praise of Slums," *Foreign Policy*, October 2012.

33. UN-HABITAT, op. cit. note 24.

34. UN-HABITAT, *Cities, Slums and the Millennium Development Goals* (Nairobi: 2006).

35. UN-HABITAT, op. cit. note 24.

36. Ibid.

U.N. Funding Increases, But Falls Short of Global Tasks (pages 117–19)

1. Global Policy Forum (GPF), "Total UN System Contributions," at www.globalpolicy.org/un-finance; U.N. General Assembly (UNGA), "Budgetary and Financial Situation of the Organizations of the United Nations System," 3 August 2010, p. 132; Tokyo Fire Department website, at www.tfd.metro.tokyo.jp/ts/sa/p03.html. The two-year 2010–11 budget runs to $4.3 billion. The 2011 figure represents half of this total.

2. New York City Office of Management and Budget, "The City of New York Executive Budget Fiscal Year 2012," May 2011, at www.nyc.gov.

3. GPF, op. cit. note 1; UNGA, op. cit. note 1.

4. Ibid.

5. Constant dollar values calculated on basis of U.S. GDP deflator series published by U.S. Department of Commerce, Bureau of Economic Analysis (BEA), "Table 1.1.9. Implicit Price Deflators for Gross Domestic Product," at www.bea.gov. BEA offers deflator values indexed to 2005 = 100; data re-based to 2011 = 100 by the authors.

6. United Nations, "Growth in United Nations Membership, 1945–Present," at www.un.org/en/members.

7. Klaus Huefner, presentation at conference on "Financing the UN: More Effective Funding for Global Priorities," New York, Global Policy Forum and the Friedrich Ebert Foundation, 11 February 2009.

8. United Nations, "Financing Peacekeeping," at www.un.org/en/peacekeeping, accessed 24 February 2012.

9. United Nations, "United Nations Peacekeeping Operations. Fact Sheet: 31 January 2012," at www.un.org/en/peacekeeping.

10. Stockholm International Peace Research Institute, "Background Paper on SIPRI Military Expenditure Data, 2010," 11 April 2011, at www.sipri.org/research/armaments/milex/factsheet2010.

11. The total number of armed forces worldwide was 20.3 million in 2010; International Institute for Strategic Studies, *The Military Balance 2010* (London: Routledge, 2009).

12. United Nations, "Structure and Organization," at www.un.org/en/aboutun/structure; U.N. Environment Programme, "United Nations Specialised Agencies versus United Nations Programmes. Note by the Executive Director," 7 June 2010, at www.rona.unep.org.

13. International Labour Organization creation from "Origins and History," at www.ilo.org.

14. Authors' calculation based on data in UNGA, op. cit. note 1. For some U.N. entities, no 2011 data were available; gaps filled by information from individual organizational websites. The $20 billion excludes funding for the International Fund for Agriculture.

15. GPF, op. cit. note 1; UNGA, op. cit. note 1; United Nations, "Peacekeeping Fact Sheet Archive," at www.un.org/en/peacekeeping.

16. Increase after inflation calculated on basis of U.S. Department of Commerce, op. cit. note 5.

17. Eva Maria Weisser, "Financing the United Nations," Friedrich Ebert Foundation, New York Office, March 2009.

18. GPF, "US vs. Total Debt to the UN: 2011," at www.globalpolicy.org/images/pdfs/US_vs._Total_Debt_May_11.pdf.

19. Calculated from UNGA, op. cit. note 1, and from GPF, op. cit. note 1.

20. Ibid.

21. U.N. Information Service Vienna, "Concern Expressed over Increasing Extrabudgetary Funding of UN Activities, as Fifth Committee Continues Debate on 2006–2007 Budget Proposal," 27 October 2005.

22. Capital Master Plan website, at www.un.org/cmp.

23. Peter Utting, "UN-Business Partnerships: Whose Agenda Counts?" United Nations Research Institute for Social Development, Geneva, 2000.

24. UNGA, op. cit. note 1.

25. Maria Ivanova, "A New Global Architecture for Sustainability Governance," in Worldwatch Institute, *State of the World 2012* (Washington, DC: Island Press, 2012), pp. 104–17.

26. Huefner, op. cit. note 7.

The Vital Signs Series

Some topics are included each year in *Vital Signs*; others are covered only in certain years. The following is a list of topics covered in *Vital Signs* thus far, with the year or years they appeared indicated in parentheses. The reference to 2006 indicates *Vital Signs 2006–2007*; 2007 refers to *Vital Signs 2007–2008*.

ENERGY AND TRANSPORTATION
Fossil Fuels
Carbon Use (1993)
Coal (1993–96, 1998, 2009, 2011)
Coal and Natural Gas Combined (2013)
Fossil Fuels Combined (1997, 1999–2003, 2005–07, 2010)
Natural Gas (1992, 1994–96, 1998, 2011–12)
Oil (1992–96, 1998, 2009, 2012–13)
Renewables, Efficiency, Other Sources
Biofuels (2005–07, 2009–12)
Biomass Energy (1999)
Combined Heat and Power (2009)
Compact Fluorescent Lamps (1993–96, 1998–2000, 2002, 2009)
Efficiency (1992, 2002, 2006)
Geothermal Power (1993, 1997)
Hydroelectric Power (1993, 1998, 2006, 2012)
Hydropower and Geothermal Combined (2013)
Nuclear Power (1992–2003, 2005–07, 2009, 2011–12)
Smart Grid (2013)
Solar Power (1992–2002, 2005–07, 2009–12)
Solar Thermal Power (2010)
Wind Power (1992–2003, 2005–07, 2009–13)
Transportation
Air Travel (1993, 1999, 2005–07, 2011)
Bicycles (1992–2003, 2005–07, 2009)
Car-sharing (2002, 2006)
Electric Cars (1997)
Gas Prices (2001)
High-Speed Raid (2012)
Motorbikes (1998)
Railroads (2002)
Urban Transportation (1999, 2001)
Vehicles (1992–2003, 2005–07, 2009–13)

ENVIRONMENT AND CLIMATE
Atmosphere and Climate
Carbon and Temperature Combined (2003, 2005–07, 2009–10)
Carbon Capture and Storage (2012–13)
Carbon Emissions (1992, 1994–2002, 2009, 2013)
CFC Production (1992–96, 1998, 2002)
Global Temperature (1992–2002)
Ozone Layer (1997, 2007)
Sea Level Rise (2003, 2011)
Weather-related Disasters (1996–2001, 2003, 2005–07, 2009–11, 2013)
Natural Resources, Animals, Plants
Amphibians (1995, 2000)
Aquatic Species (1996, 2002)
Birds (1992, 1994, 2001, 2003, 2006)
Coral Reefs (1994, 2001, 2006, 2010)
Dams (1995)
Ecosystem Conversion (1997)
Energy Productivity (1994, 2012)
Forests (1992, 1994–98, 2002, 2005–06, 2012)
Groundwater (2000, 2006)
Ice Melting (2000, 2005)
Invasive Species (2007)
Mammals (2005)
Mangroves (2006)
Marine Mammals (1993)
Organic Waste Reuse (1998)
Plant Diversity (2006)
Primates (1997)

Wars (1995, 1998–2003, 2005–07)
Small Arms (1998–99)

Reproductive Health and Women's Status

Family Planning Access (1992)
Female Education (1998)
Fertility Rates (1993)
Gender Gap (2012)
Maternal Mortality (1992, 1997, 2003)
Population Growth (1992–2003, 2005–07,
    2009–11)
Sperm Count (1999, 2007)
Violence Against Women (1996, 2002)
Women in Politics (1995, 2000)

Other Social Topics

Aging Populations (1997)
Climate Change Migration (2013)
Co-operatives (2013)
Educational Levels (2011)
Homelessness (1995)
Income Distribution or Poverty (1992, 1995,
    1997, 2002–03, 2010)

Language Extinction (1997, 2001, 2006)
Literacy (1993, 2001, 2007)
International Criminal Court (2003)
Millennium Development Goals (2005, 2007)
Nongovernmental Organizations (1999)
Orphans Due to AIDS Deaths (2003)
Prison Populations (2000)
Public Policy Networks (2005)
Quality of Life (2006)
Refugees (1993–2000, 2001, 2003, 2005)
Refugees-Environmental (2009)
Religious Environmentalism (2001)
Slums (2006)
Social Security (2001)
Sustainable Communities (2007)
Teacher Supply (2002)
Urbanization (1995–96, 1998, 2000, 2002,
    2007, 2013)
Voter Turnouts (1996, 2002)